Great Minds: Isaac Newton, Nikola Tesla, and Albert Einstein

Founders of the Scientific Age

By Mark Steinberg

©Copyright 2016 WE CANT BE BEAT LLC

Copyright 2016 by Mark Steinberg.

Published by WE CANT BE BEAT LLC

Krob817@yahoo.com

Table of Contents

Chapter 1 5
   "And there was light! 5

Chapter 2 16
   The Lone Scholar 16

Chapter 3 29
   Newton The Mighty Sorcerer 29

Chapter 4 35
   Newton Brings Light to The World 35

Chapter 5 45
   Newton's last years and the clash of the mathematicians: Newton vs. Leibniz 45

Chapter 6 53
   Boy Wonder of Lika 53

Chapter 7 66
   Nikola Tesla: "give me electrical engineering or give me death! 66

Chapter 8 83
   Tesla the Electric Wizard's Apprentice 83

Chapter 9 89
   Tesla brings more light to the city of lights 89

Chapter 10 95
   Nikola Tesla and Thomas Edison: the apprentice meets the master 95

Chapter 11	104
The Electrical Wizard's Apprentice Becomes the Master: Tesla's Return to Fame	104
Chapter 12	133
The Electric God	133
Chapter 13	149
The God of Thunder Goes to Colorado Springs	149
Chapter 14	157
A Sage is Born: The Young Albert Einstein	157
Chapter 15	165
Age of Miracles	165
Chapter 16	187
The Materialist Mystic: Einstein and the Search for Beauty in the Cosmos	187
Chapter 17	201
Sage of Princeton: Einstein comes to America	201
Chapter 18	212
Final Years of the Sage and Epilogue	212
Bibliography	217

Part 1

Isaac Newton

Chapter 1

"And there was light!"

The poet Alexander Pope once said of Isaac Newton "Nature and Nature's Laws lay hid by night: God said 'Let Newton be!' and there was light." On December 25, 1642, the momentous event took place. A child who would become one of the greatest thinkers of modern history was born: Isaac Newton. He was born in the manor house of the hamlet of Woolsthorpe which was northwest of the village of Colsterworth and south of the town of Grantham. His father, also named Isaac Newton, was the son of a yeoman who had recently become lord of a manor. His mother was Hannah Ayscough, the daughter of James Ayscough, a member of the English Gentry class. The infant was born premature and his mother joked that he could be popped into a court mug.

The elder Isaac Newton died on October 1646, two months before his son was born. Biographers of the younger Isaac Newton have noted that his father's side of the family did not have a tradition of education. Both his brother and nephew were illiterate until their deaths. The elder Isaac Newton also didn't read and write and may not have been interested in giving the young Isaac Newton the education that he received. It was his mother who came from the Gentry class who chose to send him to school to get an education. Although his father's side of the family was made up of mostly illiterate yeoman, his mother's side of the family was educated and his uncle, William Ayscough, was a reverend who had studied theology at Trinity College, Cambridge which the young Isaac Newton would later attend. If he his father had lived and had a greater role in his education, Isaac Newton may have simply continued in his ancestor's traditions and have become a farmer and we would never have heard of Isaac Newton the scientist.

In 1646, Hannah Ayscough was remarried to the Reverend Barnabas Smith. Smith did not bring the young Newton to his home and the three-year-old Isaac and the boy was left in Woolsthorpe in the care of his maternal grandmother. Isaac Newton, later in life remembered this as a very painful time in which he resented being separated from his mother. He developed bitterness towards his parents and once, out of frustration, he threatened to "burn them both and the house over them." Other than this painful experience, we do not know much of Newton's early years. In 1653, after his step father's death, his mother returned and live with him bringing with her two daughters and a son from her second marriage. Newton had trouble getting along with his stepsisters and stepbrothers and may have felt some envy towards them.

Newton's stay with his mother was brief and in 1654, she sent him to the King's Grammar School for boys in nearby Grantham. It is here that Newton first developed his intellectual interests and skill at innovation. It was also here that he met the headmaster Mr. Stokes who saw the boy's potential and played a key role in convincing his mother to allow him to stay in school to prepare for the university.

While at The King's Grammar School in Grantham, Isaac Newton learned Latin and Greek literature and the Bible which were standard for the day. He lived with a local Apothecary, whom he called Mr. Clark who had two sons and a daughter. Isaac Newton did not interact with them much. As a child, was known for being solitary, lost in his own world, and he was also a difficult and irritable child who got into fights with Clark's children over bread. He later confessed to getting into confrontations with Clark's children as well as not getting along with the other boy's at his school. While Isaac

Newton made few friends at the school, he excelled at learning. Curiously, the boy who would one day invent calculus, probably did not study any mathematics while at Grantham. He mainly studied classical Latin and Greek works which were the standard curriculum for Grammar School in those days. This was however not useless to him by any means. In the 17th century, Latin was still the language of science and all major scientific works were published in Latin. This is of course reflected in the Latin title of his principle work on physics and mathematics *Philosophae Naturalis Principia Mathematica*. It was his knowledge of the Latin language which allowed him to converse with scientists across Europe and keep up to date with the cutting edge science of the day. Poor Latin would have been much more of a barrier to his success as a scholar than a late start in mathematics, particularly since it was scarcely for years after he started his mathematical education that he invented calculus. It was also from these Classical and

Biblical sources that he most likely developed his interest in history, theology, and prophecy on which he probably spent more time than physics or mathematics.

Since he had few friends and didn't like playing games with the other boys, he appears to have spent most of his time reading and studying. He would read Greek classics such as the Ovid's Metamorphosis and Homer's Illiad. In addition to these books he also read books on natural philosophy. One book on natural philosophy which had a major and lasting influence on him was *Mysteries of Nature and Art* by John Bate. Natural philosophy was not a part of the curriculum at the Grammar School at Grantham, but Isaac Newton's curious mind led him to read many books outside of what he studied in school. This eagerness to learn greatly impressed the school headmaster Mr. Stokes though it caused Newton to be estranged from the other boys who perceived that he was very different from them. From reading books on Natural Philosophy he gained the inspiration to try to design gadgets mechanical devices. One device that he was particularly interest in was the sundial. After reading about how to measure time using shadows he made his own sundial and then proceeded to make many different dials to tell time which he left around the house of his hose in Grantham, Mr. Clark. He even went as far as drive pegs into the wall of Mr. Clark's house in order to track the daily motion of the sun and shadows created by it in order to create a clock. This probably exasperated his host. probably exasperation of his host. He also designed lanterns and even kites. This caused the other boys to think he was even stranger and avoided him. This may have been because they were not able to relate to him because he was so perceptive and cunning. There is one story where they were having a jumping contest and Isaac joined in the game. He used the trajectory of the wind to optimize his jumping distance and beat them all.

Despite the social and interpersonal struggles that accompanied his learning experience at Grantham, Isaac Newton thoroughly enjoyed studying at the Grammar School. It was here that he gained his first exposure to learning. He just could not get enough of it. The reader can then of course understand the young Isaac's dismay, when he mother called him back to Woolsthorpe to take over the estate. Since the elder Isaac Newton had been Lord of the manor, the manor rightfully went to his eldest son, that being the younger Isaac Newton. In 1659, when Isaac Newton was about seventeen, his mother made him come home to Woolsthorpe thinking it was time for him to leave school and start learning to manage an estate as its rightful heir. A servant was hired to train the young Isaac Newton in managing the land. Isaac Newton remembered this as a very difficult and depressing time. It would soon become very clear that the young Isaac Newton was not meant to be a farmer.

The servant tried to train Isaac to manage the farm, watch the sheep, and tend to the fields, but Newton would get easily distracted by other things that were on his mind. While he was supposed to be watching the sheep, he would be building windmills and designing irrigation systems. Meanwhile the sheep would go into a neighboring field and eat the neighbor's corn. The neighbors would then become angry at Newton and demand reparations which his mother would of course have to pay. When he was supposed to go to the market in Grantham with the servant to learn to sell produce and negotiate with merchants it is said that he would bribe the servant to drop him off somewhere where he would read a book or design gadgets while the servant would go to the market and pick up Isaac when he returned. When he did go to town it was to go to Mr. Clark's house where he would read and continue his old projects. As a youth he tended to be lost in thought constantly thinking about philosophical questions or problems. There is one story that he was walking his horse down a hill between Grantham and Woolsthorpe and he was so lost in thought that the horse got loose from its bridle and went home on its own without Isaac even noticing. In addition to this Isaac Newton often forgot to eat his meals because he would be busy with one of his projects or reading a book. Suffice to say, the servant became increasingly exasperated. Newton did not get along well in general with those in the household. In a list of sins he compiled later while in Cambridge, Newton confesses to having punched one of his sisters and having a falling out with the servants. Newton probably did not want to be at home and probably still felt a tinge of bitterness towards his stepsiblings since they competed for his mother's attention.

Seeing that Isaac Newton was not fit for a rural life, Mr. Stokes and Isaac's maternal uncle, Reverend William Ayscough convinced his mother to return him to Grantham so that he could finish school and go to the university which seemed to be the best place for him since he seemed be much better at dealing with ideas than well, everyday life.

Isaac Newton was overjoyed when he returned to Grantham Grammar School in 1660 after nine months at home. Mr. Stokes was very enthusiastic about continuing the boy's education and even allowed him to lodge with him and pay the extra 40-shilling fee that he had to pay to go to school at Grantham as an out-of-town boy. Isaac Newton excelled at his studies and by 1661 at the age of about eighteen he was ready to enter Cambridge. The servants at Woolsthorpe Manor were also very glad that Isaac Newton had finally returned to school considering him useless, insufferable, and in their own words, "good for nothin' but the 'Versity!"

Chapter 2

The Lone Scholar

In June of 1661, Isaac Newton began his studies at Trinity College in Cambridge. At this time Trinity College, was not known for its innovation. The curriculum was still based off of the synthesis of Aristotelian philosophy which had been forged by thinkers such as Saint Thomas Aquinas four centuries earlier when the school was founded. Though cutting edge in the 13th century, Aristotelianism was now considered a backward-looking, hidebound curriculum. Revolutionary idea from the continent however also starting to penetrate the conservativism of Cambridge University.

The neighborhood of Cambridge wasn't great either. Outside the university it was a crowded, squalid city with narrow streets. There is one story of a scholar getting his ears cut off by thieves when they found money in his gown when he had sworn that he had no more money to give them. There was also hostility between the townspeople and the university establishment which considered the university establishment to be exploitative and unfair to them. The city streets had a strange mixture of beggars, thieves, merchants, and gowned students traveling through them. It was in this environment that Isaac Newton found himself in late 1661.

Isaac Newton spent his time reading the works of modern then modern philosophers such as Rene Descartes in addition to the required reading at the university. This allowed him to keep up to date with the latest ideas in science, philosophy and mathematics. Isaac Newton was unable to receive funding and had to fund himself as a subsizar or a poor scholar who paid for his education by bringing the students food and cleaning the chamber pots among other menial tasks. Isaac Newton had hoped that life would be better at Cambridge, and it was to some degree, but he was just as isolated and found it just as hard to relate to the students at Cambridge as had been for him to relate to the boys at Grantham. Most of the wealthy students he had to serve showed little interest in learning and the other subsizars, though more serious about academics, were very narrow in their academic interests only focusing on their intended vocation. His lows status as a subsizar waiting on students only inculcated is isolation since he was isolated not just by interests and personality but also status. Isaac Newton nonetheless was able to study and was able to support his studying through being a servant to a fellow student, Babington, who was a relative of his original Grantham hosts Mr. and Mrs. Clark. By 1664 he received a scholarship and did not have to fund himself. It was also at Cambridge that he was first exposed to higher mathematics and astronomy. Although he had little mathematical education at Grantham he apparently able to quickly compensate and before long he had surpassed his teachers in his understanding of mathematics.

In 1665, the Great Plague hit London and Cambridge university was closed as a measure of caution. Many scholars were evacuated from Cambridge to escape the loss of England's intellectuals due to the plague. Isaac Newton during this time returned to his family Manor in Woolsthorpe. It is while there that Isaac Newton came up with some of his greatest discoveries, experiencing the most creative years of his life. At Woolsthorpe, he had a lot time to explore and to think about ideas which he had encountered while at Cambridge. These included the ideas Rene Descartes, Walter Charleton, Galileo Galilei, and Johannes Kepler. Now, without the distraction of classes and work he had time for his mind to roam and to explore the great questions in science of his day. Two of these major questions were the nature of light and the relation between earthly motion and celestial motion.

In the summer 1665 or 1666 Isaac Newton began to think about light much in the same way that

Albert Einstein would think about light two and a half centuries later. During Isaac Newton's time color, was thought to be an intrinsic property of objects and that light was just a substance that illuminated it. In 1664, Isaac Newton had read a book by Rene Descartes on the nature of matter and light. Rene Descartes believed that matter was continuous made of a space-filling substance and believed essentially the same thing about light. Walter Charleton on the other hand held to the view that matter was made up of indivisible corpuscles or atoms following the ideas of the Greek philosopher Epicurus who was one of the first western philosophers to propose that matter was made up of atoms. These atoms, unlike modern atoms, were simply solid particles as opposed to the modern view of the atom being composed of protons neutrons and electrons. Isaac Newton preferred the view that matter was made up of atoms or the atomistic view as it was called at the time. He applied this to light as well, believing that light was also made up of invisible light particles. Isaac Newton began to experiment with light. He experimented with prisms and noticed that if a prism was placed in a dark room and a single sunbeam was shined onto it, a white beam of light came from the prism. One-day Newton did an experiment where he set up two prisms and closed the shutters only letting a small beam of light through the window to shine on the prism. A beam of solid white light came from the first beam and a beam of many different colors came from the second beam. From this, Isaac Newton concluded that color was in fact a property of light and that white light was a combination of all the many different rays of light each one a different color. He believed that as light went through the prism it was refracted and that caused light to separate into its constituent parts. He also found that if one were to repeat the experiment with light of only one color that in both the first and second prism only one color of light would be present. Since the light came through the shutter as a solid beam, Newton imagined it as a solid beam of particles confirming in his mind that light was particulate. Through this experiment, Isaac Newton overturned a long standing theory going back to the philosophers of ancient Greece that light was a simple substance. That is not bad for a twenty-three-year-old genius.

Optics was not the only field Newton was revolutionizing while essentially vacationing at Woolsthorpe Manor. He was also thinking about astronomy. Galileo, in addition to providing empirical evidence that earth and the planets orbited the sun and not vice versa, also came up with the theory kinematics which describes the motion of objects on earth. Johannes Kepler come up with the three laws of planetary motion which described motion of the celestial bodies. Thus Galileo had described earthly motion while Johannes Kepler had described celestial motion. At this time, though there was no common theory to explain all motion. The heavens and the earth seemed distinct, ruled by different laws. Rene Descartes had also suggested that the planets of the solar system were caught in a vortex made up of multiple layers and that each of planets and comets were embedded in a particular layer of vortex and moved with the vortex. Although this would explain heavenly, or celestial, motion, it did not explain earthly motion This disunity between the heavens and the earth was one of the issues which bothered Isaac Newton. The mystery of what made things move was a question which went all the way back to the Pre-Socratic philosophers Heraclitus, Parmenides, and Democritus who all had different explanations for motion. The most popular one in Europe until the end of the Middle Ages was of course that of Aristotle who explained motion in terms teleology. He believed that different things moved to where they were supposed to go. A rock fell to the ground because that was simply where it belonged, for example. Aristotle also talked that matter was inherently lazy and preferred to be at rest. If you pushed a cart, it would inevitably return to being at rest. To Aristotle rest seemed to be the natural or most basic state of matter. This was the prevailing view in Medieval Europe, but the Middle Ages were over and Aristotle's ideas were no longer sacrosanct having been challenged by recent scientists such as Nicholas Copernicus of course the aforementioned Galileo Galilei and Johannes Kepler. Isaac Newton as he often did, solved a two-thousand-year puzzle during his brief hiatus from Cambridge in 1665-1666 and again in 1666-1667.

One day in 1666, as the story goes, Isaac Newton was sitting overlooking his orchards with the moon visible in the sky. He was thinking about planetary motion when he saw an apple fall from the tree. This caused him to think about the motion of the apple in relation to the motion of the moon. The apple had to have accelerated since it went from not moving to falling as it fell from the tree. He reasoned that there must be a force drawn the apple to the ground. The force attracting the apple to the ground must extend at least to the original height of the apple. At this point Newton came to a realization. Why should the force stop at the height of the apple tree? After all, if the tree were on top of a mountain, much higher than the tree, the same thing would happen. This force must then extend to even the tallest mountains, but why should stop there? It is at this point that he came to one of his most famous, powerful ideas, the theory of universal gravitation. Why couldn't the force that he saw acting on the apple extend all the way to the moon? In fact, what if the moon itself was held in its orbit around the earth by the same force that pulled the apple from the tree branch to the ground? Over the next few years Newton began to develop the mathematical basis for the theory and eventually came to the familiar equation.

$$F = \frac{G m_1 m_2}{r^{\wedge}2}$$

G is the gravitational constant discovered later by Henry Cavendish in 1798, m1 and m2 are the masses of two objects, r is the distance between their centers, and F is the gravitational force between them. What is so fundamental about this theory is that it created a unified theory of motion. Before this the universe had been separated into celestial and earthly domains with separate laws governing both. The universe was hierarchical with separate spheres. The theory of gravity said that both the celestial motion of the planets and the earthly motion of apples and cannonballs were essentially the same type of motion just on a different scale. The celestial realm was in fact no different from the earthly realm in terms of physics. Gravity is also what made Isaac Newton the first astrophysicist since he was the first to use physics to explain astronomical phenomena. Before this astronomy had merely been a branch of mathematics.

Isaac Newton, in order to show that his theory was valid, had to provide a mathematical basis for this. He started out with geometry and algebra, but he soon realized that the math of his day was not advanced enough to handle such a problem. He realized that he needed a way to express rates of change as things that were themselves constantly changing. This required calculus, so using geometry and algebra he invented calculus to evaluate his theory. This shows the genius of Isaac Newton, when was the last time you just invented a branch of mathematics because you couldn't figure out a math or physics problem with the math that you had available?

Isaac Newton returned to Cambridge briefly in 1665 to receive his B.A. and returned again to receive his M.A. in 1667. In 1667 he also was given a position as a lecturer at Trinity College. He lectured on mathematics and optics. He was finally being paid to do something that made good use of his skills, well, to some degree anyways. Newton was definitely brilliant but he was not the greatest lecturer that Cambridge had ever seen. It is said that very few attended his lectures and even fewer actually understood them. He would lecture on every day, except for Sunday out of deep religious convictions, never taking a break. On days where lectures were not held he would simply lecture to the walls. Sadly, the walls may have been his most devoted students. Despite his mediocre lecturing he was still awarded the Lucasian chair of mathematics in 1669 when Isaac Barrow retired.

Chapter 3

Newton The Mighty Sorcerer

Though he made remarkable achievements during the time he was away from Cambridge between 1665 and 1667, he did not initially share his ideas, possibly because he was afraid of how other would respond to his work. Isaac Newton is known to have been quite paranoid and may have been afraid that they would try to steal his ideas or harm him in some way. He was also very sensitive to criticism and may have feared a critical response. During the 1670s his interests shifted away from physics and astronomy and towards chemistry and alchemy. What many people do not know about Isaac Newton is that he wrote more about alchemy, mysticism, and religious texts such as the bible, mainly the Bible actually, than he did about physics, mathematics, or astronomy. Isaac Newton believed that the ancients had discovered the inner laws governing the universe and the one could gain scientific insights into nature through studying their writings. Newton was also deeply religious and believed that all insights into nature revealed the nature of the Creator. Newton's ultimate goal was to discover how God created the universe and through that to understand God. In his alchemical studies he searched for the philosopher's stone which was believed to turn base metals into gold and also the elixir of eternal life, but alchemy is ultimately a religious endeavor, a quest for the divine. His study science and his occult studies were simply two different sides of the same search.

In addition to alchemy, Isaac Newton also tried to predict the end of the world. He studied the books of Revelation and Daniel in the Bible to try to establish when Christ would return. He predicted it would not happen before 2060. Newton didn't see himself as a prophet of doom however. In his writing, Newton explained that his motivation for calculating the timing of the apocalypse was to put on end to all the speculation and numerous claims about the end of the world during his time. Similar to America and to some extent Europe today, Europe of the seventeenth and eighteenth centuries was full of doomsday prophets such as Nostradamus who predicted the end of the world coming in a matter of years. Newton, annoyed with all of these popular claims, did calculations which he believed showed that the earliest date the world could possibly end according to the scriptures was 2060 and that that he could not know for certain that it would even end on that date. He did however know that it would not end before that date, rendering all the seventeenth and eighteenth century claims that it would end in a few years debunked. Isaac Newton was in fact trying to deter end times hysteria rather than perpetuate it. Although Newton certainly had many mystical beliefs, he was still a skeptical individual who did not buy into anything that he heard unless he believed there was sound reasoning and evidence behind it. He was still a rationalist. Isaac Newton also wrote a great deal on biblical chronology and tried to reconcile ancient Greek stories with the Biblical account of history. What is interesting is that Newton appears to have believed not only in the historicity of the battle of Troy, that it was a historical event that happened as described in the Iliad, but also that the voyage of Jason and the Argonauts was historical since the date of the voyage is one of the anchor points in his chronology. He placed the voyage of Argonauts and the battle of Troy, which took place shortly after the voyage, around 936 BC. This is five hundred years later than the date given by most classical historians.

The reason that Newton dated the events this way is because he was trying to match historical events with past astronomical alignments which could be calculated from the regular motion of the celestial bodies. Using these astronomical alignments and the number of years recorded in biblical and classical texts, he believed that he could create a chronology more accurate than the other chronologies being produced at the time such as the famous biblical chronology produced by James Ussher in 1650 known for marking October 23, 4004 BC as the date of creation of the world. The question of whether Isaac Newton succeeded in creating an accurate chronology I will leave to historians to slug it out over in their debates. The point is that Isaac Newton was a very prolific man who wrote on a great many topics, from mathematics to the dimensions of the Temple built by King Solomon.

Many people are surprised to learn that Sir Isaac Newton, often thought of as a great scientist and rationalist, was interested in the magic and the occult. It must be remembered thought that in the seventeenth century magic and science were not as distinct as they are now. Many scientists were involved in what would be considered magic today and some magicians did things that would now be considered science. Johannes Kepler for example worked as an astrologer for the king of Denmark before working with Tycho

Brahe. John Dee, the famous agent of Queen Elizabeth I known for his accomplishments as a mathematician and astronomer was also a sorcerer and diviner who invented an alphabet to communicate angels. Isaac Newton's esoteric interests were typical for scientists of his day. At the same time, it was also in the seventeenth century that magic and science began to diverge. Robert Boyle laid down the foundations of modern chemistry beginning its divergence from alchemy and Isaac Newton himself was responsible for the astronomy becoming more about studying the physical universe and less about divination and predicting the future.

Chapter 4

Newton Brings Light to The World

In 1679, Isaac Newton's attention was returned to astronomy and physics after engaging in correspondence with the astronomer Edmund Halley and the appearance of the comet which bears his name around 1681. In 1679, Isaac Newton was contacted through letters by Robert Hooke secretary of the Royal Society. Robert Hooke was responsible for keeping correspondence with various royal society members learning about what sort of research was being done by them. Isaac Newton in his letter response described his theory of universal gravitation and the idea that gravity was governed by an inverse square law. The inverse square law says that if you have two bodies and you double the distance between them, the gravitation attraction is decreases to a quarter. If you half the distance between them, it increases by a factor of four. This was the first time that Newton began to share his ideas with the wider scientific community beyond just his close friend. It unfortunately was not a pleasant experience for Newton since his ideas were not met with immediate acceptance. Many scientists criticized his theories on light because they failed to replicate his experiments exactly. There also established figures within seventeenth century scientific figures who felt their prominence threatened by this young scientist and philosopher from Cambridge with his new, groundbreaking ideas. One of these scientists was Robert Hooke who was seven years older than Newton and claimed that he had already come up with the idea of the inverse square law before Newton. This began one of many disputes or "science wars" if you will, that Newton had with other scientists over his ideas. Newton, who did not like criticism, found the process very unpleasant. He was taken aback by the hostile response to his ideas and was resentful of Hooke's claims that the inverse square law of gravity was not his own. Newton's pride was so hurt by his dispute with Hooke that he did not publish his ideas until after Hooke's death so that Hooke could not respond to Newton's refutation of his claims.

The reason for the science war between Newton and Hooke goes back to the fact that around the same time that Newton was changing history with his experiments and mathematical work at Woolsthorpe in 1665 and 1666, Robert Hooke independently came up the idea of a mysterious force which attracted the sun and planets to each other and that it varied with distance between the two bodies. It should be noted however that Robert Hooke did not explicitly say that the variation with distance was governed by an inverse square law. He also did not provide rigorous mathematical proofs. Nonetheless, he did come to a conclusion very similar to Newton's independently.

Coming back to Isaac Newton, Newton circulated his writings among his close associates, but he did not publish them because of the unpleasant experience he had the first time he tried to go public with his ideas in 1679. His friends nonetheless encouraged him to publish his work before someone else published the idea. During this time, he expanded on his theory of motion and developed his famous three laws of planetary motion.

1. An object in motion or at rest will remain in motion or at rest until acted upon by an external force
2. When an object is acted on by an external force, the force exerted on the object is directly proportional to the mass of the object and inversely proportional to its acceleration. This is described by the equation $F = ma$ where m is the mass of the object, a is its acceleration, and F is its force.
3. With every action there must always oppose an equal and opposite reaction

These still apply today to all structures larger than the atom from dust particles to galaxies and are the basis for most modern mechanical technology. When we get to Einstein we will talk more about the cases where these laws begin to break down.

Newton was once again at work making new discoveries in mathematics, physics, and astronomy and in 1686, with contributions from the astronomer-royal John Flamsteed, he published the first volume of the book *Philosophae Naturalis Principia Mathematica* which laid down the mathematical foundations of his contribution to physics. The second volume was published in July 5, 1687. It is at this point that controversy started earnest because on the publication of the Principia Mathematica, Robert Hooke insisted that he had come up with the idea of the inverse square law and gravity and that Newton had borrowed his ideas during their letter correspondence in 1679. This began a long, bitter feud between Newton and Hooke which continued until Hooke's death in 1703. Isaac Newton insisted that he had not borrowed from Hooke and that the idea of the inverse square law was entirely his own. He also added that even if he had borrowed from Hooke that he still deserved some credit since he made the mathematical proofs supporting the theory. Eventually, Newton ended up winning out but still gave some credit to Hooke saying that Hooke had indeed independently come up with the idea around the same time.

After the publication of the *Principia*, Newton became a renowned scientist across Europe and was hailed as the founder of a new science. Isaac Newton, by establishing a set of mathematical and physical laws that described all motion, made it possible for the entire universe to be understood in a mechanistic way like a giant clock operating according to internal mechanistic patterns. Different parts of the universe become like different gears within the clock. This idea that the physical universe could be understood as giant machine like a clock helped to launch the Industrial Revolution of the late 1700s. If the behavior of matter could be understood in a mechanistic way like a machine, then this knowledge could be used to manipulate matter to create smaller machines within the larger cosmic machine. It would simply amount to a rearranging of the gears. Furthermore, if the behavior of the physical universe could be described like a machine such as mechanical clock, then why not everything else in reality including human behavior and social institutions? This is what later thinkers during the eighteenth century Enlightenment would do, or at least try to do. Adam Smith attempted to describe the economy in a scientific or Newtonian way by formulating laws which governed the markets such as the law of supply and demand and the invisible hand as if the economy itself was a machine. You could say that "Newtonianism" led to capitalism. It is simply looking at society in the same mechanistic way that scientists look at the universe.

Anthropologists Lewis Henry Morgan and John Stewart Mill in the 19th century would do the same with human society in general describing laws which governed the evolution of human societies from simple to complex. Even political and social institutions could now be described as machines where humans were just cogs in the machine thanks to Newton.

Human society was just another machine which operated by laws which could be determined through reason, experimentation, and observation. Not only that, but human society would later be seen as a consequence of the of the physical universe. The physical bodies interacting with each other are ultimately responsible for human society and human thought. This is still a major tenant of philosophical naturalism today, the idea that everything in reality including culture and consciousness has a material cause deriving from physical matter. Isaac Newton would have not agreed with this idea. He was no materialist, but he was inadvertently responsible for making materialism a more attractive option. To be fair, this idea of a mechanistic clockwork universe does not belong to Newton alone. Rene Descartes also contributed to the idea that the universe was a machine which was governed by deterministic laws, but it was Newton who provided a mathematical and experimental basis for a mechanistic clockwork universe, allowing for it to become a widespread understanding of the world. Descartes also was quite clear that there was also a non-material component of reality and did not believe that knowledge of the material universe was sufficient to fully understand the nature of reality. Newton would have agreed with Descartes on this point, but Newtonian mechanics seemed so complete in its description of the universe that later naturalists became convinced that because of it, knowledge of the material universe was all that was needed to understand the full nature of reality. These later thinkers essentially launched the Enlightenment. It is significant to realize that Newtonian mechanics provided the basis for both the Industrial Revolution and the capitalism of Adam Smith. If you think about this, you will realize that anyone who lives in an industrial, capitalist society with rockets, airplanes, and satellites is at least partially indebted to Isaac Newton because all of these things are dependent on a Newtonian conception of the universe in some way or another.

Chapter 5

Newton's last years and the clash of the mathematicians: Newton vs. Leibniz

Isaac Newton, although somewhat more well-known because of the *Principia*, was still very isolated as he had been all of his life and still tended to work on his own. Nonetheless, by the 1690s he was a respected scholar and a successful scientist with a fruitful career on which to look back. He also continued his side projects such as studying biblical prophecy and looking for the philosopher's stone. All however was not well. In 1692, Isaac Newton suffered a nervous breakdown, the cause of which is still not entirely understood. During this time, Isaac Newton became even more paranoid then usual accusing his friends of trying to undermine him and steal his work without any evidence. He also terminated on long term relationship a friend he had known since his early days as a student at Cambridge and wrote a harsh letter to his friend Edmund Halley. He also appears to have suffered from hallucinations and heard sounds that were not there. This episode lasted several months. Some scholars have suggested that he experienced mercury poisoning from his alchemical experiments while others suggest that the cause of his nervous breakdown was years of isolation and working without almost any time off. Those who knew him say that they never saw him going out to a tavern or engaging in any form of recreation. Isaac Newton was almost always in his study and library pouring over books on natural philosophy, mathematics, theology, and ancient occult wisdom among other things, though his friends only really knew about his studies of mathematics and natural philosophy.

After recovering from his nervous breakdown, he began to become more interested in politics. Isaac Newton, being from a university which received support from the crown, was a royalist and a staunch opponent of Catholic intrusion into England. In 1696, after thirty-five years at

Cambridge, he finally left the city to go to London to become master of the mint. Isaac Newton the scientist was now Isaac Newton the administrator who oversaw the coining of British currency, caught counterfeiters and assayed gold and silver. He also pursued his political career with the same zeal with which he had pursued science. In 1703, Robert Hooke, one of Isaac Newton's major rivals finally died. Isaac Newton, back in the 1670s said that he would prefer to wait for his older rivals to die before he published all of his mathematical works. Well, Robert Hooke was finally dead, so at the prompting of the mathematician John Wallis, Isaac Newton began to work on his last major scientific work. In 1704, he published *Opticks* in which he outlined his mathematical ideas in a way which, surprising for Newton, was actually easy for most people who were knowledgeable of the subject to understand. The book quickly gained him widespread fame across Europe. It was during this part of his life that Isaac Newton became a celebrity known by all scientists across

Europe. On the continent he was known as one of the greatest British natural philosophers. In Britain he was talked about almost as if he was immortal. Isaac Newton for the last twenty-three years of his life was a household name among scientists, philosophers, kings, government officials, and most educated people. He was the Stephen Hawking of the early eighteenth century. This brought Newton much fame, but it also, as was often the case with Newton, brought controversy.

In *Opticks*, Newton made it clear that he was the inventor of calculus which was at the time considered to be the invention of the German mathematician Gottfried Leibniz. Isaac Newton had developed calculus in 1666 while in Woolsthorpe Manor during the plague years. He however did not publish anything on calculus, or *fluxions* as he called it, until 1693. Gottfried Leibniz had begun to independently develop a theory and notation for calculus in 1676 and first published his work in 1684. Historians now know that Leibniz and Newton both independently arrived at calculus. Isaac Newton did not see things this way at the time though and insisted that Leibniz had stolen his ideas. Now these were both well-known and respected intellectuals and both had a lot of supporters across Europe, though most of Newton's supporters lived in Britain. This would be like Stephen Hawking and Roger Penrose accusing each other of stealing the other's ideas. Scientists and mathematicians across Europe were divided into two camps. Continentals supported Leibniz and wrote many letters present what they viewed as evidence that Leibniz was the true father of calculus. Meanwhile the supporters of Newton did the same or him. This long drawn out battle outlasted the death of Leibniz in 1716, and even after his death, Newton was still creating works which showed that he was the true originator of calculus. In a way both Newton and Leibniz won and they both lost. Newton was hailed as the true inventor of calculus by the time of his death but it is Leibniz' notation and terminology in calculus that is most widely used rather than Newton's notation. This is mainly because Newton did calculus in a way that was much more confusing and hard to understand. After Newton's death, continental Europe used Leibniz notation while British scientists stuck to Newton's notation for a century before adopting Leibniz notation. The use of Newton's convoluted notation actually slowed progress in British mathematics considerably.

Before moving on let's briefly discuss what this last idea that Newton introduced to mathematics is exactly. For those who have not studied calculus, calculus is essentially the study of quantities that are constantly changing. For a simple example, suppose you were filling a bucket with water and you wanted to know the time it took to fill the bucket. If the rate at which the water fills the bucket is constant, then you can simply divide the total amount of water in the bucket when it is full by the rate at which you filled the bucket, but what if the rate at which the bucket is filling is constantly changing? In this case you would need to use calculus because the rate of change is not constant. Calculus basically measures quantities that are constantly changing. This is enormously helpful for science since there are many scenarios where rates are of change are complex and require calculus to get right. If we did not have calculus, figuring out things like how to launch a rocket would be considerably more difficult.

Other than his spat with Leibniz, Newton spent the rest of his life basking in popularity. In 1716, he published another addition of the *Principia* and he also chose to publish some of his works on biblical prophecy and chronology. These were some of the concluding thoughts of this genius before he died on March 20, 1727.

There you have it, Newton gave us gravity, the laws of motion, a clockwork machine universe, and calculus. He also made contributions to the field of optics. These are all foundational to our modern world. Most of our non-electrical technology is still based off of Newton's laws of motion and gravity, the idea of clockwork machine universe permeates western philosophy, and where would we be without calculus? This concludes our discussion of Sir Isaac Newton, oh yes almost forgot, after publishing *Opticks* in 1704 he was knighted by queen Ann. That is right, Isaac Newton was a scientist, a sorcerer, and a *knight*. Think about that about that for a while. We are now going to move on to the nineteenth century early twentieth century to consider yet another great mind whose contributions are of course indebted to Newton.

Part II

Nikola Tesla

Chapter 6

Boy Wonder of Lika

Nikola Tesla is one of the least appreciated great minds of the nineteenth and twentieth centuries. Although he turned away rewards and money for his work, he is responsible for the age of electricity when electricity is such a regular part of everyday life that it fades into the background. We use electric lighting, electric household appliances, electric cars, and almost every advanced piece of technology that we use has an electrical component of some sort these days. The reason I am sitting at this laptop writing this book is in part thanks to Nikola Tesla. Very few however know this. The reason for this is that Tesla, much like Newton was a very private individual. He avoided friendships, romance, pleasure, and any kind of connection that would separate him from his work. As an inventor, he saw himself as being an instrument to be utilized for the progress of mankind. Because of this, he had to focus all of his energy and passion into invention and use of the mind. He could not have any unnecessary attachments because that would distract him from invention and would be a disservice to mankind. He believed that the best way to make himself an efficient and productive inventor was isolation. He was never married and died penniless since all of his money went into his inventions. He did not spend any money on his own leisure, though he did take money to support his basic needs. In a way, he was successful. His inventions and ideas have brought us an age of electric power which has allowed for mass production of goods and a high standard of living at least in the first world. One the other hand his isolation caused him to be unsuccessful in that many of his ideas were not fully implemented. After his death, most of his work fell into obscurity and most people forgot about the boy wonder of Lika and the far seeing inventor and engineer that he became. Recently interest in his work is being revived. Nikola's goal was ultimately to bring about an age of prosperity, peace, and progress for humanity and perhaps if Tesla had done more to connect with other humans, we might be closer to that goal. Now, let us explore the ideas and life of this fascinating individual.

The great inventor came into the world on midnight July 9 or in the morning hours of July 10, 1856 in the village of Smiljan in the Lika province of the Austro-Hungarian Empire to the Serbian Orthodox priest Milutin Tesla through his wife Djouka Tesla. Milutin had been a soldier and had been educated in a military school, but left the military to become a priest and poet and married a daughter of priest. Djouka, though she could not read or write was a brilliant woman who was capable of memorizing Serbian epic poems and passages from the Bible. She was also well known for her skills with a needle and thread. Nikola would later say that he was indebted to his mother's ancestry for his mechanical proclivities. Even from a young age Nikola was perceptive. When he was less than five he would wander the mountains around Smiljan catching frogs and trying to fish. One time he made a crude windmill while in the mountains near his home and another time, even though he had not used or even seen one, he fashioned a wooden fishing hook and attempted to use it to fish. He made it after hearing about how a group of his friends had gone fishing without him. Tesla would later recollect that he enjoyed being close to nature all those years ago in hills of Smiljan in the country of the Serbs.

Around 1863, when Nikola was five, Milutin Tesla was assigned to pastor a church in the town of Gospic. When they moved to Gospic, Nikola had a lot of trouble adapting to his new environment. In Smiljan, he had spent much of his time alone wandering the hills and valleys. He had been close to nature. The new artificial atmosphere of Gospic with its relatively crowded streets and an environment where he had to constantly interact with other humans made Nikola long for rustic and quiet Smiljan. Tesla was a shy child and stayed away from the boys and girls of the town. The girls were beautiful and hard for him to approach, the boys were all taller and stronger than him. He was thankful when his father gave him the job of ringing the church bell on Sundays to call people to mass since it meant that he could stay out of sight. One time, a wealthy matron with a long train extending from her skirt had come to visit the church. As she was leaving Tesla tripped and fell on the train causing it to come off. The woman was furious and his father was angry. Tesla was an embarrassment to the entire town after this, and he would not be able to exonerate himself from this status until an incident occurred which foreshadowed his later mechanical genius.

One day, a water pump was being installed for the fire department and the entire town was gathered to watch its demonstration. Tesla watched, fascinated, and wanted to get closer as did most of the other children. As they were watching it, the firemen had trouble getting the pump to work. As they were trying to fix the problem, Telsa spontaneously got the idea to go check the part of the pump that was taking water from the nearby river. He jumped into the river, swam up to the pump and realized that one of the valves had a kink in it which was obstructing the water flow, he fixed it and the pump started working perfectly. After this everyone was very impressed and they were finally able to forget the unfortunate incident at the church with the matron. Tesla later recalled that he had no idea how the machine worked. He just had the intuition to check pump. This was only the beginning of the flourishing of Tesla's exceptional and even strange mechanical abilities.

At about the age of twelve, Tesla attended the Real Gymnasium in Gospic. Here he studied hand drawing and mathematics. Tesla did not like hand-drawing as he was left-handed which left him at a disadvantage. He received the lowest grades in his hand-drawing class. This however was not because he was unable to do it. He could have gotten better grades in the class, but there was one student who was not a very good artist who was striving to do well in hand-drawing so he could obtain a scholarship. He would not be able to get a scholarship if he received the lowest grades, so Tesla, wanting to help his classmate, intentionally got the lowest grades possible. Tesla does always seem to have been in some ways self-denying and altruistic to an extreme degree. This is reflected in his later years when he took it upon himself focus entirely on invention for the good of humanity. Whether or not this is truly altruism I will leave to the philosophers. I will simply say that it was much more altruistic than those who made inventions simply for the sake of become famous or wealthy. Tesla seems to have gone out of his way avoid either rewards of fame or fortune seeing himself as only an intellectual servant of mankind. This was certainly altruistic, whether this also made him humble or selfless is a topic for a different book.

The subject in which Tesla most excelled was naturally mathematics. He had the ability to picture problems so clearly in his mind that he was able to solve the problem before he had time to draw it out on a blackboard. His teachers at first thought that he was cheating, but they eventually realized that he had a mental gift. The strangeness of this gift does not stop there. He had the ability to picture objects in his mind so clearly that is was sometimes hard for him to tell whether he was actually seeing it or if it was just a mental picture. This was a nuisance for his everyday life but was of great help while studying math. The subjects he enjoyed most were however not a part of the school curriculum. He enjoyed reading the books in his father's study and would spend hours after school reading them. His father at one point forbade him from doing so because he was afraid the boy was going to ruin his eyes from reading in the dim candlelight. Nikola started taking books and candles up to room to continue reading which worked for a while until he got caught. It seemed nothing could deter the young Nikola from quenching his thirst for knowledge.

Although Tesla had an interest in mechanical gadgets, the school had nothing to offer in terms of classes at Tesla's level. Tesla ended up reading books on mechanical gadgets in his father's study. Also on his own time, he engaged in metalworking and woodworking outside of classes, applying his skill in building things. In addition to having an aptitude for mathematics he was also skilled at learning foreign languages. Nikola during his time at the Real Gymnasium learned French, German, and Italian. This allowed him access to far more literature than was available to his classmates who only spoke his native Serbo-Croat tongue. One day, the Real Gymnasium had an exhibition on waterwheels which were used to produce electricity from water. Nikola was inspired by it because he like machines. He had also seen pictures of Niagara Falls. He thought of how much power he could generate if he were to build such a machine at Niagara Falls. It is at this point that he purportedly told his father that one day he would go to America create power from Niagara Falls. That would however have to wait many years, about thirty years actually.

There is another story of Tesla while he was in the Real Gymnasium at Gospic that he attempted to create a flying machine. Somehow Tesla learned that all objects on earth had pressure of 14 pounds per square inch from the atmosphere bearing down on them. He also learned that in a vacuum, there is no air pressure. From this, he got the idea to make a contraption where two halves of a cylinder were attached, one filled with air and the other a vacuum. His idea was that the pressure in one of half of the cylinder would cause the other half of the cylinder to move and the cylinder would begin to rotate. He believed that this would make the cylinder rotate fast enough that all he had to do was attach a propeller and he had a flying machine. He then just needed to strap it to his back and he could fly, oh the ideas that come from the mind of a twelve-year-old boy. He created the machine and placed it in a box and the cylinder did spin but only very slowly then it stopped. Distraught he double checked everything to see what had gone wrong. Eventually he came to the conclusion that he was not able to keep air from leaking into the vacuum. He also would later realize that he was making incorrect assumptions about the direction in which the air pressure was exerting force on the cylinder. He thought that it was tangent or parallel to the surface of the cylinder when it was in fact normal or perpendicular, that is, that the air pushes against the walls of the cylinder rather than circulating within the cylinder. This was not a total waste though, and he would in later years use this concept in creating a very power-efficient steam driven turbine called appropriately enough the "Tesla turbine."

One important thing to know about Tesla as a boy is that he always thought in very big terms. He thought of everything on the grandest scale possible. One day while in the mountains he noticed lightning hit the ground and only moments later it began to rain. From this Tesla concluded that lightning caused the rain. This was incorrect but it led him to the conviction that electricity could be used to control the weather. If the weather could be controlled by electricity, then he could harness electricity to make it rain in deserts and increase the world's food supply so that there was no shortage of food in the entire globe, putting an end to famine. Had he known about them at the time, he probably also would have believed that it could be used to break up hurricanes and other destructive storms before they caused their damage. Year later he would attempt to use electricity to do just that, control weather. He did not succeed because rain does not work that way, but it illustrates the nature of his thinking. To Tesla, there were no limits to what humans could do once they learned to harness the power of nature. He saw infinite progress from use of the mind. His belief in the power of the mind is exemplified by the next chapter in his life.

Chapter 7

Nikola Tesla: "give me electrical engineering or give me death!

Nikola Tesla finished his schooling at the Real Gymnasium by 1870. He was fourteen at the time. One of his first tasks upon graduation appears to have been to organize and classify the books at the school library. Tesla was delighted at fist in this endeavor, but after a while became ill, most likely from overwork and malnutrition. Tesla worked very hard on everything he did and tended neglect things like eating and sleeping. The consequence of this was a bout of poor health on occasion. Tesla's condition continued to worsen until he was no longer able to work on the books and became bed-ridden. Nikola Telsa's father Milutin was very concerned and believed that his son Nikola was a very delicate boy. Milutin was afraid that his son with his tendency to work without any breaks would work himself to death. Having already lost a son, Milutin Tesla wanted to do everything he could to preserve his son's health even if that meant not allowing him to go into a technical field which Milutin knew would require a lot of effort and studying on Tesla's part. Milutin Tesla thought that it would be better for Nikola if he went into ministry and joined the priesthood. If he became a priest, he would not have to study as long and the work would be less demanding. Nikola continued to wrestle with the illness and began to despair of life. He recounts that what finally caused him to gain the willpower to overcome his illness was when he looked from his sickbed at a copy of a book by Mark Twain on his bookshelf. This somehow gave him hope and inspired him to keep battling the illness and live. Later, after immigrating to the United States, he would meet Mark Twain and they would become good friends, most likely because of this.

In 1871, Nikola Tesla left Gospic to attend the Higher Real Gymnasium in Karlovac. The Higher Real Gymnasium was something between a high school and a junior college within the academic program that existed in the Austro-Hungarian Empire at the time. Tesla did not entirely enjoy his time at Karlovac. His aunt with, whom he lodged while living in the city, would limit how much she fed him believing that feeding him too much would harm his frail constitution and "overload his stomach." He was not fed much meat as a result. He was also frequently sick from malaria which he caught from the lowlands. You could say that not all was well with our young inventor. He did however have some positive experiences. One such experience helped him to decide his future career path. He met a physics teacher who was known for his experiments. He dazzled the young Nikola Tesla with all sorts of electrical experiments. Electricity at this time was mysterious and not well understood. As a result, it was often the subject of spectacles and it was very mysterious almost magical to non-scientists. I suppose you could compare it to how quantum mechanics or the more esoteric aspects of relativity such as wormholes are viewed nowadays, very weird. It was after this experience that Nikola Tesla decided his future profession. From that time onwards he wanted to be an electrical engineer. He finished at the Higher Real Gymnasium in 1873 at the age of seventeen and headed home to tell his parents the news that he had chosen to study electricity and use it to harness the power of nature and bring about human progress.

When he returned home however, all was not well. His parents had actually written him a letter asking him not to come home because of a cholera epidemic that was currently ravaging Gospic. They didn't actually say this in the letter though, they simply told him to go on a long hiking trip to regain his health. Nikola Tesla who didn't get the letter and came nonetheless. His parents were dismayed that he with his poor health had come during a cholera epidemic. They, particularly his father, were even more dismayed when he said that he wanted to become an electrical engineer. When Nikola Telsa learned of his father's plans for his future education and career, he was devastated. Not only did he have to go into ministry which he had no interest in doing, he also would have to serve three years of compulsory military service. He had wanted to immediately start at an engineering school and start studying the mysteries of electricity. In addition to this, on the same day that he got back and heard the news, he contracted cholera. He became very ill and was bed-ridden for nine months. His father, desperately not wanting to lose another son did everything he could to save Nikola, everything except let him become an electrical engineer, at first. As his condition continued to worsen A feeling of apathy came over Nikola. If he lived he would have to go through, what was to him, the excruciating affair of three years of military service and then train for, what was to him, a dreary life in the priesthood, not to say that being a priest is inherently dreary of course. Many people lead very happy fulfilling lives as members of the clergy; this would not have been the case for Tesla though. Both possibilities, death on one hand and compulsory military service and life in ministry which he did not want on the other seemed equally unpleasant to him. At this point Nikola did not care whether he lived or died. One day his father came to his sickbed, urging his son to find the will to live. His son told him that he would only be able to get well if his father let him become an electrical engineer. Milutin Tesla, being a desperate father who only wanted his son to live, finally conceded and insisted that he only wanted his son to be an engineer. Nikola Tesla at this moment had the will to overcome his illness and live.

Nikola Tesla recovered fairly rapidly in response to this. Within a few days he was able to sit up in bed and in a few more days he was already able to walk. He was very excited to start engineering school. There was however one problem. Nikola being a subject of the Austro-Hungarian Empire was still required to serve for three years in the military as a soldier. His father, trying to find a solution, told Nikola to "go on a hunting trip to recover his health."

While Nikola Tesla hid in the mountains for a year, Milutin Tesla looked for a solution trying to convince friends and relatives in the military to find a way to exempt Telsa from military service on account of his poor health and other factors which he could bring up as excuses. Meanwhile, Nikola Tesla, who was just waiting up in the mountains had a lot of time to think and to imagine. During his time in the mountains he spent his time thinking of vast engineering projects to implement one day. One idea was an underground tunnel connecting Europe and the United States to deliver mail. Another project he thought of was an enormous solid ring encircling the earth at the equator above the surface. He imagined it being in orbit and being used as a form of rapid intercontinental transportation. He did not actually have a plan for implementing most of these ideas. Most of them were just for fun to pass the time. He explored the physics, mathematics, and engineering concepts involved. This is the kind of young man that Nikola Tesla was, the kind, who to pass the time, designed globe-spanning, civilization-shaping engineering projects. Later he would use ideas and mathematics from these projects in his real world projects such as the building of the Tesla Turbine.

In 1875, Nikola Tesla resumed his education at Graz, Austria at the Graz Polytechnic Institute. He was nineteen at this point. Ready to dive into his dream career he began to plan his life, though he had not planned it fully at this point. He thought of his life as simply another engineering project. His main goal was to maximize efficiency and output. As an electrical engineer he would devote his life entirely to science and to the improvement of mankind. To be as efficient and productive as possible as an electrical engineer, he removed any chance for romance or recreation from his life and devoted himself entirely to his studies so that he could do what he needed in order to bring many inventions as possible to the human race and thus improve the human condition. He would only sleep from 11 PM to 4 AM and spend almost the entire rest of the time on his studies even on holidays and weekends, at least Isaac Newton, being religious, took the day off on Sundays. Tesla would only take a break to sleep for a few hours and even then he would read himself to sleep. He did this partly for the reason described and also partly as a way of showing his appreciation to his father for letting him attend engineering school by working as hard as possible. At the end of the fall term he received the highest grades possible in all of his classes and took twice the number of subjects requires. The dean of his department said that he was "as star of first rank" in a letter to his father.

Upon returning home after the first term, Nikola Tesla excitedly told his parents of the grades and honors he had received for the work he had done. His parents however only seemed somewhat interested in his grades and were much more concerned about his health and his father scolded him for taking such a risk with overwork and urged him to drop out for the sake of his physical well-being. Nikola would later learn that his father had earlier received a letter from the school advising that Nikola be withdrawn from the school out of concern that he would kill himself from overwork.

During his next year, Nikola Tesla did less work and took fewer classes, mainly physics and mathematics. Tesla was also exposed to the Gramme machine during his second year which would lead to one of his greatest inventions, the alternating current polyphase motor. It was also at this time that he found himself in the midst of a dispute with one of the professors, Professor Poeschl, over the viability of one of his ideas. The Gramme machine was a machine which could operate according to a battery or an electrical dynamo. It was a direct current device and used a commutator to maintain a direct current. Telsa was amazed when it was first displayed at Gratz by Professor Poeschl. While watching it, spellbound, he noticed that it sparked quite a bit and he asked why. The professor told him it was because of the commutator. Tesla at this point suggested that a way to improve the efficiency of the device would be to use an alternating current which would eliminate the need for a commutator. The professor disagreed which resulted in a long though friendly argument between instructor and student over the soundness of the idea. The next class time Professor Poeschl actually took a break from his usual lecturing schedule to address Tesla's idea in front of the class. After what he considered to be a crushing critique of Tesla's idea, the professor said "Mr. Tesla will accomplish great things but he certainly will never do this." This did slow the young Tesla down a little but as he thought about the professor's case against alternating current, he realized that all the professor had really done was to argue that no one knew how to do it and not that it could not be done. After this realization, Tesla spent most of the rest of the second year trying to produce a machine that ran on alternating current. He spent more time working on this project than he spent working on his studies This was possible because he had done so well he first year that he did not need to focus as much on his studies for the second year. Toward the end of the second term however, he began to despair that he was going to accomplish his goal of finishing his project, vindicating himself after he had been publically discredited in front of the class by Professor Poeschl. He did not succeed in making an alternating current machine by the end of term but it did lay the groundwork for later successes. The effort of trying to get his alternating current motor to work culminated in him eventually finding the solution six year later while going for a walk, but we are getting ahead of ourselves.

During Tesla's second year, for whatever reason he began to focus less on his studies and engage in more pastimes. One of these pastimes was playing cards and billiards. He enjoyed cards since the game was analytical with a lot of mathematical patterns and gave his mind many problems to solve. His analytical mind made him quite good at the game. Whenever he made a winning he would always give the money that he had won back to the loser. His opponents were unfortunately not as altruistic. This led to trouble when, while back in Gospic for the summer, he gambled and lost money that he was supposed to have used to cover traveling expenses for a trip to Prague as well as his tuition money. He was unable to get it back the way that those who lost games to him were able to get it back. When he confessed this to his mother, she, rather than being upset that he had lost the money, was hopeful that running out of money for school would compel him to turn to a profession that put less of a strain on his health. She even gave him more money for gambling and simply said "Here, satisfy yourself."

Nikola Tesla played again and was fortunately able to win the money back and this time he did not return it. After this, Tesla vowed never again to gamble later saying "I conquered my passion then and there and tore it from my heart so as to leave no trace of desire."

Nikola Tesla had studied so little because of his gambling the previous year that at the start of his third year, he felt academically unprepared. He asked the school for an extension, but the extension was denied. After this, the young Tesla was forced to drop out of Graz School of Technology. Records indicate that it may have been because of his gambling and claims of him being a "womanizer." Nikola Tesla was afraid to tell his parents, most likely out of shame and embarrassment, and chose to disappear without telling them where he was or what had happened. He friends searched for him everywhere and failed to find him. They eventually figured that he had drowned in a nearby river.

Nikola wandered through nearby towns leaving Austria into Slovenia. Eventually in early 1878 and came to the town of Maribor where he got a job as an engineer making 60 florins a month. He did not have the job for long however and soon he was once again unemployed. He also began to gamble again playing cards with men on the street. In 1879, Milutin Tesla learned that his son was alive from a cousin and went to Maribor to find his son. Finding Nikola Tesla he urged him to return to Graz. When Tesla refused and explained the situation, his father offered to pay for him to attend another university, the University of Prague. Tesla agreed and returned home to Gospic. During the time between his father finding him and the time he went to Prague, Nikola Tesla stayed with his family reliving some of his younger years before he had left to the Real Higher Gymnasium and the Graz School of Technology. During this time, he apparently met a girl named Anna who, if Tesla had not still been intent on his goal of becoming a celibate inventor and automaton produced for the advancement of the human species, may have been a potential love interest. It is told that they would go for walks through the nearby wilderness and that at one point, Tesla event told her that he loved her. This would represent a rare moment in Nikola's life in which he allowed himself to consider romance an acceptable option. Nevertheless, their goals went in different directions and such a union was not to be. Nikola Tesla wanted to become an inventor, and Anna wanted to start a family. Upon going to the University of Prague Nikola Tesla, continued his study of physics and mathematics. While there he also met the philosopher Carl Stumpf. Stumpf was an empiricist who believed that humans were born with a blank slate, essentially without any innate knowledge and that everything they learned was from the senses. Nikola Tesla also read on Descartes who believed that animals were automatons, mere biological machines part of a larger cosmic machine, an idea which also comes from the cosmology that western science, at this point in time, had inherited from Sir Isaac Newton. This idea of the universe as one vast machine would influence Tesla' worldview as he came to view the entire universe as a machine, a machine that could be manipulated for the benefit of the human race.

In 1880, Nikola Tesla's father Milutin died. Without his father to fund his education, Tesla now had to be self-supporting. He graduated from the University of Prague and made his way to Budapest where he learned that a central telephone station had just been established. Nikola Tesla went there with the hope of finally attaining his goal of working as an electrical engineer.

Chapter 8

Tesla the Electric Wizard's Apprentice
Hoping to find a job, in 1880, the twenty-four-year-old Nikola Tesla headed off to to the city of Budapest to find a job at central station that was to be built for telephone installations. When he arrived there though, he soon realized that he would not be able to get a job yet because the enterprise had not yet begun and was still in the planning state. Nikola Tesla was without money and still needed to find a job quickly. He was only able to get a job as a draftsman at the central telegraph service run by the government. The pay was meager but it was something and it was at least related to what Tesla wanted to do. As he worked, the master-in-chief of the service soon noticed his skill and gave him more and more responsibilities. When the telephone station was put into service in 1881, Tesla was placed in charge. At last Tesla was where he wanted to be. He was a professional electrical

engineer and could resume his, so far, life-long desire to build an alternating current machine.

One thing we have learned about Tesla is that he worked very hard. His time at Budapest was no exception. He still only gave himself five hours of sleep and typically only slept for about two of them. This had over the years taken a toll on them and he reached a breaking point in 1881 when he suffered a strange mental breakdown. Tesla suddenly became hypersensitive to everything around him. A feather hitting the ground sounded like a bomb exploding. He also could feel the presence of objects at a distance from an odd, "creepy" feeling on his forehead. He also felt tremors all over his body. To this day, we do not know what happened to Tesla only that no one could do anything to help him and that he was forced to take time off work until he recovered. It is told that he was near death during this time but what kept him alive was his burning desire to find a way to make the alternating current machine work. Ever since he had the humiliating experience of the Professor Poeschl telling him that his idea was unfeasible and that he would never succeed in front of his entire class no doubt, Tesla had a desire to prove the professor wrong and that an alternating current machine was possible, one that would obviate the need for a commuter and would be much more efficient. The machine would make work more efficiently and would make life easier for workers. Tesla resolved that he had to do it. He had to succeed or he would die. Luckily, Tesla did overcome his mental condition, whatever it was, and was able to return to work by the beginning of 1882. It was during this year that he found the solution to the problem which had made him unable to make an alternating current machine.

One day he and a friend were walking through a park in Budapest when Tesla had a realization. Staring at the setting sun he had one of those very clear images in his mind that he had experienced since childhood, an image so real and solid it was hard for him to distinguish it from reality. He pictured a machine with an alternating magnetic field and a stationary circuit which would drive the alternating current. He, according to his friend, stood motionless for a minute or two staring at the sunset saying "Watch me! Watch me reverse it!"

His friend for a moment thought that he was talking about attempting to reverse the course of the setting sun and supposed that his friend had been under a bit of stress lately. He even asked if Tesla was ill. Tesla then turned to him and explained his idea. The two apparently talked about it most of the night. While they were walking Tesla took a twig and began to draw a diagram onto the dirt path on which they were walking to illustrate. He had already thought it through enough in that short time to draw a diagram. Attempts at creating alternating current engines had been made in the past but so far they had been done in such a way that the current could not be sustained because it would change as fast as the current itself. Rather than creating a driving force, the machine would just create oscillations which did not contribute to the work done by the machine. Tesla used two circuits with the same alternating current where the waves were out of phase with each other. In addition to figuring out how to create an alternating current engine, the creation of a rotating magnetic field in of itself was an impressive accomplishment. Prior to this time, electrical engineers had essentially used to a stationary magnetic field to spin a series of coils. Tesla made instead a magnetic field that itself rotated so that it did not require the spinning coils. This was superior since it did not require wires to keep the current going through the coils, simply the rotating magnetic field.

During this time Tesla went through something of what Isaac Newton went through in the summers of 1665 and 1666. Since he was creating all of these machines in his imagination he could make it however he wanted and was not limited by raw materials. Most of the later machines that he would design, he came up with during the short period between the time that he had the experience and the time that he left Budapest in early 1882. Tesla used his unique visualization skills to design many different motors which used single phase, double phase, and multi-phase alternating current. He was happy that he had last solved a great scientific problem of his and could focus on creating one that was more than just a mental model.

Chapter 9

Tesla brings more light to the city of lights The same year that Tesla conceived the rotating magnetic field, the central telephone station in Budapest where he worked closed and he was encouraged by the family friend who had originally found him a job in Budapest, Ferenc Puskas, to go to Paris to work for Edison Continental, the continental European branch of the Edison company. Tesla was excited about the prospect of working in Paris. It was an international and cosmopolitan city which at the time was a major center of innovation. Telsa hoped that Paris could be place where he could bring his idea of the alternating current machine to the entire world. After his arrival, he was involved in many projects. He helped set up the lights in the Paris theater and numerous street lights. To Tesla's frustration, they all used direct current and he is recorded as having felt it quite cumbersome to have to use the direct current motors when the alternating current motors were so much more efficient in his eyes. His manager was the colleague of Thomas Edison, Charles Bachelor. He was one of the men who worked on the first phonograph and is known to have worked on the filament for the lightbulb. Charles Bachelor was open to new ideas but appears to have been ultimately more interested in profit. Something that Tesla noticed a lot in his industry of choice. During his days in Paris he would get up at five in the morning to go for a swim in the Sein River and then would walk to work, presumably after changing his clothes. In the evening, he would play billiards with the men of Paris. Tesla loved to tell others about his idea of the alternating current. People must have become tired of it at times, and that appears to have been the case. Of all the people he talked to, no one seemed to be that interested in his ideas. Even Mr. Cunningham, a good friend whom Tesla had met in Paris who was from America, was only interested in the possibility of starting a stock company through the invention. This is one thing that is notable about Tesla, his unfailing altruism when it came to his inventions. He was not interested in recognition or in profit beyond what was necessary for his livelihood. He simply wanted to make inventions that would improve the human condition. While most inventors would have been very secretive about their discoveries, fearing their ideas might be stolen, Tesla talked about his idea to everyone he met. It is unfortunate that few seemed to want to listen.

Over the course of the next two years that Tesla worked at Edison Continental in Paris, he worked essentially as what one of his biographers calls a "trouble-shooter." He would travel around France and Germany responding to problems or malfunctions in electrical installations which had been set up by the company. Around April of 1883, Tesla was sent to the city of Strasbourg to respond to a serious problem which threatened to cost the company a lot of money. During an opening ceremony in Strasbourg for the first electrical generator during which, Emperor William I was present, a short circuit due to defective wiring caused the generator to fail. The German government in response threatened to not allow the company to continue its operations in the town. Edison Continental, afraid of losing their investment, sent Tesla to deal with the situation. Tesla saw this mishap as his opportunity to try to test his alternating current motor with a rotating magnetic field by building one to replace the defunct one in Strasbourg. He began to work on the alternating current using his ability to visualize models. Because of this, he was apparently able to build the machine almost exactly as he had planned. He did keep a notebook of his designs and probably used it as an aid. Using the notebook and his brain, he was able to design the machine exactly as he had originally pictured it. The moment of truth came when he flipped the switch to activate the AC current machine. It worked. After seven years he had finally exonerated himself of the crushing critique administered by professor Poeschl. He had finally done it. He was right. His idea worked. After the demonstration was a success, Tesla's work caught the attention of the mayor of Strasbourg, Mr. Bauzin. Mr. Bauzin was enthusiastic about Tesla's work while Edison Continental seems to have been ultimately uninterested in Tesla's innovation. Mr. Bauzin offered to help Tesla promote the AC current and start a company which sold it in Strasbourg. In order to accomplish this, the mayor tried to interest the wealthy men of Strasbourg. He gathered them and Tesla demonstrated is machine. Bauzin and Tesla were unfortunately unable to garner their support however, and Mr. Bauzin eventually told Tesla that he would be better off returning to Paris where his ideas might garner more interest. In spring of 1884, Tesla returned to the city of lights hoping that they might listen. While in Paris, he also hoped to gain compensation for his work in Strasbourg. His compensation was however meager and far below the $25,000 that he thought it should be worth considering all of the work that he had done in Strasbourg. Feeling swindled, Tesla decided to leave Paris and go to America at the encouragement of Charles Bachelor to work for Thomas Edison himself. It is possible that he hoped that working directly with Edison would allow him to work on more innovation and not just be a repairman for motors.

Chapter 10

Nikola Tesla and Thomas Edison: the apprentice meets the master

Upon arriving in New York, Nikola Tesla said that he considered the continent to be uncivilized and backwards. Life in America he thought was like life in Europe a century before his time. By that, he probably meant the relative lack of cutting edge technology like electric lighting in America versus Europe where it had become more common. There is one story about Tesla when he first came to New York where he was walking down a street and came across a machine shop. In the machine shop he noticed that a man was working on an electric machine. Walking into the shop, he asked about the machine and the man said he had done all he could to try to fix it without success. Tesla then offered to fix it without charge. There is also a version of the story where he asks the man what he will give for twenty dollars. Whatever version of the story is correct, Tesla certainly made his mark on America and would spend most of the rest of his life in the country. When he met Edison at his laboratory at Goerck Street, which had once been an ironworks, he was, on one hand, dazzled by the man. Here stood the inventor of devices such an electric pen, a musical telephone, the phonograph, and an incandescent light bulb. He was clearly an intelligent and accomplished man who had come up with many more inventions than Tesla. There was a reason that Thomas Edison was called the Wizard of Menlo Park. Tesla later recollected that he felt like he had wasted his life in comparison to Edison in that, by comparison, he had only made a few real inventions in his life so far. One the other hand, he was repulsed by Edison. He recounts in some stories that he thought that Edison had terrible hygiene. Edison and Tesla also just didn't get along in many ways. Edison, when he first met Tesla, is purported to have asked him if he had tasted human flesh, possibly referencing some of Tesla's ancestors' traditions. Serbia is not far from Transylvania, the home of Vlad Dracula. Tesla is reported to have replied "No, and how about you?" Edison Replied by saying "I eat Welsh Rabbit" and claimed it was because it refueled his mental capacities. Tesla thought that was gross.

As can be seen, from the very start, the two men did not quite get along. Thomas Edison was an inventor and a businessman. He relied primarily on ingenuity as well as trial and error in order to accomplish his inventions as well as a savvy understanding of capitalism which allowed him to become as wealthy as he became. Tesla, on the other hand, preferred a theoretical and mathematical approach and was as interested in ideas as he was interested in inventions. He also didn't care much about the financial side of things. This gave the two men very different approaches to inventions. Tesla used mathematics and theory to guide his inventive powers whereas Edison simply used trial and error. Tesla found Edison's approach cumbersome and impractical seeing all the work Edison had to do when a little bit of math and physics would rapidly accelerate the process. When Tesla attempted to explain his idea for an alternating current motor, the idea went straight over Edison's head. It didn't help that many of Edison's competitors such as George Westinghouse and Elihu Thomson were looking into the possibility of alternating current motors, making the idea of alternating currents a threat to the success of Edison's business since all of his electrical products were direct current. Edison for whatever reason stuck to direct current. He was convinced that with a little bit of ingenuity and inventiveness one could overcome all of the problems with direct current and that anything was possible using direct current. To be fair, he had been able to come up with a great deal of inventions that were all direct current. This however did not change the fact that direct current was still very inefficient due to needing a commutator. This would be later realized during the "war of the currents" in which alternating current would win out over direct current because of its greater efficiency and, as a result, its lower cost.

Edison did not implement Tesla's idea of switching to AC current but gave Tesla a project to redesign the DC systems to make it more efficient while maintaining DC. He offered Tesla $50,000 if he was able to redesign it. Tesla agreed. While working on the project, Tesla worked hard as he always did. He tended to work from 10:30 AM to 5 AM the next day. During this time, he would work on redesigning Edison's DC motor and occasionally do on-call repair jobs. Over the course of the year that he worked for Edison, he completed 24 dynamos and was able to replace the core magnets shorter cores which were more efficient. He also got paid for installing incandescent lightbulbs and arc lights. It was not all work however and Tesla and the others who worked with Edison did still have good times. They would meet at a restaurant across the street from Edison's lab where they would tell stories and jokes while drinking. Tesla would also engage in playing billiards as he always liked to do while telling the players about his ideas, mostly his idea of AC motor and its applications.

In early 1885, Tesla finished his project of redesigning the DC motor and asked for the 50,000 dollars which he had been promised. In response, Thomas Edison laughed and said "Tesla, you don't understand American Humor." The realization that all of his overwork, all of his effort, all of the days he had not taken time off had been all for nothing and that Edison saw it as little more than a practical joke was too much for our genius inventor. Edison did offer him a $10 raise for his work, but Tesla, feeling swindled, immediately resigned. This was the beginning of the growing rivalry between Tesla and Edison who would become the two great techno-wizards of their day. This rivalry would last until the end of their lives. I suppose you could say that the apprentice had rebelled against the master.

Although it was hard for Tesla, sensitive as he was, to accept the slight, he did learn something valuable from the experience, that his rival electric wizard was not actually as formidable as he had thought. Tesla, Tesla himself thought, in fact had a better invention than Edison himself and could easily surpass Edison in prominence as an inventor. Who needed the Wizard of Menlo Park anyways? In spring of 1885, Tesla worked with Lemuel Serrel, a patent Lawyer to obtain his first patent, an arc-light that was more effective at lighting and did not flicker or spark like most arc-lights of the day. This invention appears to have drawn the attention of two New Jersey Businessmen who approached Tesla after he made his first patent in the summer months of 1885. The two businessmen expressed a vague interest in Tesla's alternating current motor and agreed to help him start a company in his name, the Tesla Electric Light & Manufacturing

Company. Tesla obtained stock in the company through purchasing and was eventually able to make enough income to buy a nice apartment in Manhattan with a garden. The first project involved installing the arc lights which he had invented on several city blocks in New Jersey. It took about a year to finish the project. Upon finishing the project, Tesla expected them to allow him to start developing and installing motors based off of alternating current since they had expressed interest in the idea in the beginning.

To Tesla's shock and dismay however, his business partners now took no interest in his alternating current idea. Not only that, but they even ended up pushing him out of the company. In the process, Tesla was left with very little money. He could not even get money from his patents because they were bound up in the company and thus royalties from the patents went directly to the company and not to him. Over the course of the year, things got progressively worse for the young inventor until late 1886 when he was out of money and was unable to find any work and was forced to work as a ditch digger for $2 a day. He later described it as a time of "terrible headaches and bitter tears, my suffering being intensified by my material want." This was potentially the lowest point in Tesla's life. He had been betrayed by people that he had trusted, he had lost everything financially, and now he was forced to be a day laborer, something he considered beneath him. He began to have doubts about the worth of his education. What was the value of it if after all those years of studying science, mathematics, and literature he was now digging ditches? You could say that at this point, it looked as if Thomas Edison had won the war of the Electrical Wizards. Edison was wealthy, famous, and respected for his inventions; Nikola Tesla was destitute, obscure, and shoveling dirt.

Chapter 11

The Electrical Wizard's Apprentice Becomes the Master: Tesla's Return to Fame

From the winter of 1886 to the spring of 1887, Tesla worked as a manual laborer and did odd electrical repair jobs when he could get the chance. During this time, he was dejected but not completely silent. He began to talk to the foreman who had hired him as a laborer and told him of his past as an engineer and inventor. The foreman ended up being impressed by Tesla's accomplishments and introduced him to an engineer who had worked with the Western Union Telegraph Company, Alfred S. Brown. Brown was equally impressed by Tesla and agreed with him about the inefficiency of direct current motors and saw value in Tesla's idea of using alternating current. Brown got Tesla in contact with an attorney by the name of Charles F. Peck. Coming to Charles Peck he explained his situation. Peck was sympathetic but unsure of how to help him. The story goes that Charles

Peck asked Tesla to give him a reason for why he should help him. Tesla, inspired by a story about Christopher Columbus balancing an egg, went to the local blacksmith to make a contraption that would allow him cause an egg to stand using alternating currents generated by a rotating magnetic field. He constructed the device and then demonstrated it to Brown and Peck. Both men were won over and the three men, Tesla, Brown, and Peck agreed to start a company in Tesla's name. This time it was called the Tesla Electric Company. They agreed to split the patents and shared the patent for the AC (alternating current) dynamo as well as numerous other patents that Tesla would make during his time with the company. It was through this enterprise that Tesla would finally attain his goal of being a career inventor. For the next decade or two he would continuously produce new inventions which would transform the world.

As usual, Tesla began to work very hard, taking very little rest. This was a pattern that he would continue all of his life. He and his colleagues at the Tesla Electric Company got to work on the AC inductor and other devices that could be made with alternating current using Tesla's design. Charles F. Peck ensured legal and financial support while Alfred S. Brown assisted with the technical aspects of the work. Another inventor by the name of Anthony Szigeti also joined them and became a direct assistant to Nikola Tesla. All these men eventually became famous as background supporters of Tesla who produced most of the inventions. He made AC systems of all kinds, single-phase, double-phase, and multiphase. He also made a combination of this. His inventions eventually gained the attention of T.C. Martin, the writer for the magazine *Electrical World*. T.C. Martin asked him to write an article in his magazine about his inventions. It did take some convincing on the part of T.C. Martin who had to enlist the help of Cornell professor William Anthony. Anthony came to Tesla's lab to test the AC dynamo himself. When Professor Anthony did the test himself, he was very impressed and, in collaboration T.C. Martin, encouraged Tesla to present it before the American Institute of Electrical Engineers (AIEE). Tesla was very much an isolated individual and preferred to work alone. Tesla, however conceded and agreed to present his ideas to the AIEE. This helped Tesla introduce himself to the greater electrical engineering community in the United States.

On May 15, 1888, Tesla gave a lecture at the AIEE. This lecture is considered very significant in the history of the field. In the lecture, he addressed what was a major controversy at the time. During this time a raging competition occurred between three rival electrical companies, that of Thomas Edison, Westinghouse Electric Company, and the Thomson-Houston Electric Company. While Edison had stuck fiercely to direct current systems, Westinghouse had begun to explore the use of alternating current or AC systems. So far, Westinghouse's engineers had been able to create AC generators which were able to extend voltages across three quarters of a mile, a little farther than the best DC (direct current) generators. The high voltage required for Westinghouse's AC systems were nonetheless considered to be very dangerous by leading electrical engineers which was pointed out by Thomas Edison in a letter in which he purportedly wrote "Just as certain as death, Westinghouse will kill a costumer within six months after he puts in a system of any size."

Tesla's idea of an AC generator with a rotating magnetic field was both safer, more efficient, and could extend a voltage much farther than any previous AC or DC system yet invented. It could be used not only for lighting but also for household appliances. After Tesla finished his lecture, Professor Anthony conceded and reported the positive results he had gained with the AC generator invented by Tesla. This revelation was not without controversy. Several other inventors had also invented various AC electric motors and they insisted that Tesla's work though independently produced was not the only approach to using alternating currents in electric power generation. The engineer Elihu Thomson, right after Tesla's lecture, got up and essentially said that although Tesla had produced an intriguing device that he had done the same earlier and so that Tesla was not the earliest originator or the AC current motor. Tesla conceded but pointed out that his motor was still more efficient in that his did not require a commutator to maintain the alternating current whereas Thomson's did. Tesla had still produced a superior motor and was worthy of the credit. Tesla' work also caught the attention of George Westinghouse. Westinghouse, another nineteenth century business tycoon and inventor known for being quite large a large man with a walrus mustache, had worked with two other engineers, William Stanley and Oliver Shallenburger who had both produced AC generators very similar to Tesla's design but they did not know the theory behind it. This put them at a disadvantage.

Westinghouse knew that there was profit to be gained from Tesla's invention and after the speech he approached Tesla with an offer to partner with Tesla's company and provide him with royalties in return for a share in the profits of his invention. At the time, Tesla and his cohort, the engineer Alfred Brown and attorney and financial officer Charles Peck, were also considering another offer from a Mr. Butterworth in San Francisco for $250,000 and royalties of $2.50 per wat of power. This was considered very high but Westinghouse realized the value of Tesla's patent. So far Westinghouse's engineers had not been able to produce one which matched the efficiency of Tesla's machine. William Stanley claimed to have come up with essentially the same design and to have written it down in a notebook in 1883. The problem was that Stanley's AC motor still required a commutator and thus was still less efficient. Westinghouse agreed to give the same amount in royalties even though he thought that they were excessive. Westinghouse would become a major backer of Tesla and play a major role in what would be known as the war of the currents.

Tesla's invention of the alternating current led to a revolution in the use of electrical power within a matter of years. All of today's electrical motors and grids are based on Tesla's alternating current idea. It can be said with certainty that Tesla was on his way to winning the war of the currents. The wizard of 38 Liberty Street, the address of Tesla's laboratory while at the Tesla Electric Company, would soon triumph over the wizard of Menlo Park. The apprentice had become the master, Gandalf had beat Saruman, or the other way around depending on whose side you take.

In addition to the $2.50 royalty on every watt of power generated by the polyphase AC motors developed by Tesla, his company was paid about

$70,000 a year in funding for his research and inventions as well as 10,000 for every patent. At this point Tesla moved from his apartment in Manhattan to live in a hotel in Pittsburg. The hotel expenses were also paid by Westinghouse who would continue to support Tesla in one way or another for the next decade. Although Tesla received funding for research he did not take any monthly or weekly compensations and received for himself about $100,000 from 1889 when he began to receive funding from Westinghouse until 1897. During this time, Tesla gained many friends and admirers such as Charles Scott, a young man who became his lab assistant and a major supporter Tesla's cause to bring the AC polyphase motor to the world, and Albert Shmidt with whom he shared patents. He did also gain many enemies who attempted to thwart his progress. Oliver Shallenburger, most likely jealous of Tesla's recognition as inventor of the AC polyphase induction motor, attempted to show that he was the real inventor of the motor with the help of his associate Lewis B. Stilwell who would later write a book on the history of the Westinghouse Electric Company. Another opponent of Tesla and Westinghouse was the engineer William Stanley who separated from the Westinghouse company around 1892 or 1893 to sell his own AC motors. His motors later turned out to essentially be copies of Tesla's motor and he was became legally required to buy the motors from the Westinghouse Electric Company. While darts were being hurled at Tesla and his inventions, there was meanwhile a larger battle being waged between two giants of the electric industry, Thomas Edison, the old Wizard of Menlo Park, and George Westinghouse. Thomas Edison still defended the DC motor of his invention and continued in his efforts to undermine the use of Tesla's AC polyphaser induction motor.

This rivalry reached a high point, or especially low point depending on how you look at it in 1889 when Edison hired an engineer Harold Pitney Brown who had an interest in using electrocution on animals for scientific study to work in his laboratory. H.P. Brown tested the DC and AC motors and found that while that while the DC current took a long time to kill the animals through electrocution, the AC current killed them almost instantly. By Experimenting on stray dogs, Brown found that a very high voltage was required to kill the dog and even at that high voltage the animal was in a lot of pain. Brown found that the AC current was relatively quick and painless though it required a much higher voltage. He realized that a motor run on alternating current could conceivably be used as a reliable and humane means of execution. At the same time H.P. Brown noticed numerous casualties within the Edison company from electric shocks, most of them were due to direct current but some were due to alternating current and Brown however focused his attention on deaths involving alternating current and decided that alternating current was too dangerous for use outside of execution. Meanwhile, he planned on designing an electric chair and patenting it.

He also purchased one of Tesla's motors to use in his experiments and began to test it on larger animals to make sure that it was humane enough for the use of human execution. Edison, seeing that this was good propaganda to use against Westinghouse and Tesla's AC induction motor, began to indirectly spread the propaganda. When it came time to test the electric chair, the subject chosen was William Kemmler, a lunatic who had butchered his mistress with an axe. During the trial of William Kemmler, experts on electricity including Edison himself were brought in to testify to the effectiveness of the electric chair. Edison, wanting to further damage Westinghouse's reputation said that the AC current was incredibly deadly and that it was prefect for execution. He also ensured them that all of his own electrical inventions were safe and not nearly as dangerous as those developed by his rival.

This decidedly harmed Westinghouse's reputation and made some customers much less enthusiastic about purchasing Tesla's equipment to say the least. Articles began to appear in periodicals talking about "electrical executioners" spreading panic and fear regarding the use of this new form of execution. It also did not help that that when the execution took place, it was decidedly not humane. A black cap was attached to Kemmler's head and when the current was applied it caused him to heave, convulse and foam at the mouth. This spectacle resulted in controversy and outrage over the messiness of the execution. One eyewitness told the *New York Times* that he would rather see ten hangings than one electrical execution. The experiment with electrical execution was widely denounced as barbaric. Thomas Edison of course encouraged this outrage, though he didn't directly propagate it. He had hired another engineer to assist H.P. Brown in his experiments and did mention in an interview how he, in reading about the execution said it was "not pleasant reading."

This controversy threatened to damage Westinghouse's profit from the AC motors because so many people, now shocked and horrified by the execution of Kemmler and the alleged dangerousness of the AC induction motor now were much less inclined to install it in their city streets and in their homes. Tesla knew that if they were to survive as a company they would have to produce motors at lower frequency and emphasize the safety of the motors. Tesla's work was not fast enough and by late 1889, Westinghouse said he could no longer fund Tesla because his motors were such bad publicity. Tesla however negotiated a deal that he agreed to discontinue the royalties from the watts of power produced by the AC induction motors. After this Tesla returned to his laboratory in New York. At this time, he still felt pretty confident. He had accomplished one of his adolescent goals, to build the AC motor and to give it to the world. He had said that he would do this back in 1876 and now thirteen years later he had accomplished it even though his own professor said it could never be done. After returning to his laboratory in New York City, he began to investigate how he could improve upon his motor. In order to do to do this, Tesla attempted to replicate the experiments of Heinrich Hertz who was studying electromagnetic waves and frequencies. This was useful to Tesla because, he needed to study the range of frequencies at which he could set his motor. After returning to his laboratory, he visited Europe and met some of his old friends as well as the physicist Wilhelm Bjerknes. He also visited his family in eastern Europe before returning to New York. The simplicity of his travels in Europe has been compared by his biographers to the much more publicized and extravagant trip taken by Edison at the same time as he was touring Europe while showing off his inventions, meeting king Humbert of Italy and major scientists such as Hermann von Helmholtz and Louis Pasteur. The bad publicity and reduced funding after the loss of royalties caused a relative stall in Tesla's work on the AC motor, but his motor was still gaining widespread use throughout the world. While he was in Europe, Swiss and German scientists had started to implement it as well as his former rivals William Stanley and Louis Stilwell.

In the same year that Tesla was in Europe, a central power station which used alternating current with Tesla's model was used to power the city of London at a distance of seven miles and across 11,000 volts. Edison, alas, was only able to hold back the rise of alternating current for so long and by 1890, the Edison General Electric Company itself was switching to AC current. Tesla worked in his own lab at this point though he also did what he could to help Westinghouse who would still turn out to be a long term ally of his whom he would later praise as one of the few men capable of "overcoming the power of money" to foist his invention onto the world which was Tesla's intention. It was not too long before most electric companies were using Tesla's design for the AC induction motor. This made Tesla very confident and he began to make increasingly extravagant claims about his inventive capabilities. One claim he made was that he could "place 100,000 horsepower" on a single wire. He also promised wireless energy which he planned to make available across the globe. These extravagant claims sometimes hurt his credibility since it caused him to be compared to sharlatans of the day who made similarly extravagant claims such as John E.W. Keely who claimed to have created a motor that he called the "hydropneumatic pulsating vacuo engine" which he claimed was run on etheric energy. Helena Blavatsky, claimed that he had discovered how to harness vryl, what she believed to be the life energy of the universe. He was able to dazzle and draw enormous crowds with this remarkable "invention." It was later discovered after his death that the source of energy for his "etheric engine" was a hidden chamber of compressed air. Like Tesla and Edison, Keely was compared to a god or demigod by his supporters. Tesla's extravagant claims and impressive inventions also fit into the general atmosphere of the 1890s. During the 1890s, it seemed like every new invention would change society forever. Optimists in both scientific and mystical circles believed that humanity was on an endless upward ascent towards progress. In a few decades all the secrets of the universe would be known and all humanity's ills would be gone forever. Nikola Tesla's claims that his inventions would alleviate the burden of work as well as make life easier and more fulfilling for all people was very much a reflection of that very optimistic era which continued until 1914 when the First World War shattered the naïve notions of progress that had so entranced thinkers of the 19th century. It could be said though that the World War I did not dampen Nikola Tesla's spirits who continued to make optimistic and extravagant predictions even after the end of the First World War.

In his new lab in New York Tesla purportedly began most of the projects on which he would work for the next half a century. His new lab was near at Bleecker Street not far from one of the laboratories of Thomas Edison. During this time, Tesla used the money he had gained from his supporters to live in fancy hotels such as the five story Astor House. Tesla was dedicated to his mission to start a new technological age. He spent most of his time in the lab while occasionally having dinner with friends. He mostly worked alone and often times did most of his work at night when there were not as many distractions. Alfred Brown would occasionally assist him but for the most part the reclusive inventor worked by himself and preferred to do so. Because of his wealth, he sent money back to his mother and sisters and would write letters to them from time to time telling them of what was happening with his life in America. He would send them about 150 florins a month or 6 months in wages for a typical Serbian worker. Nikola Tesla had become something of legend among his family and was already considered a Serbian national hero. In the year 1890, T.C. Hall convinced Tesla to publish his ideas about electromagnetic waves and their relation to light in the *Electrical World*. Tesla also told him about his childhood and his Serbian heritage of which he appears to have been proud. He told Hall about how the Serbs had bravely fought relentlessly against the invading Turks. During this same year he also met the Serbian physicist Michael Pupin with whom he worked briefly before bitterly separating after Pupin fraternized with Tesla's long-time rival Elihu Thomson which Tesla considered to be an act betrayal.

Through 1891 Tesla continued to make new inventions such as a mechanical oscillator which was able to produce a steady current and deliver up to several million volts. He also began to correspond with British scientists such as J.J. Thomson and Sir. William Crookes who were doing pioneering work in electromagnetism and waves at the time. Tesla found their work very interesting and began to correspond with them. Wanting to present them with his ideas, Tesla increased his workload and tried to repeat their experiments. During this he also invented what would eventually become known as the tesla coil. In January 1892, Tesla was invited by his old friend T.C. Martin and several other British figures in science to speak at the Royal Society. He arrived in London on January 26, 1892 and stayed in the home of Sir William Preece a very prominent old figure in British science at the time. Preece had been involved in the science of electricity and electromagnetic transmission since 1860. He had known the inventor Alexander Graham Bell and also had met Edison in 1877. Tesla was honored to be in the presence of Preece, somewhat of a giant figure in British science. During his stay, he also had the opportunity to meet Lord Rayleigh and Lord Kelvin. Before the Royal society, he demonstrated his tesla coils and the power that they could produce. He also performed other impressive feat where he allowed current to pass through his body and light two lightbulbs which he held in his hands. What Tesla was most interested in presenting though were his ideas for wireless communication. Tesla believed that a method could be attained where by power and information could be transferred through the air without the need for wires. This is one of the main ideas that he presented before the Royalty society. One of the devices he presented was an invention called a button lamp which was able to vaporize objects. The button lamp was a spherical lamp with a reflective inner surface. An object would be placed inside and then light would be emitted from the interior of the lamp which would then continuously reflect off of the reflective inner surface. Eventually this focused light would vaporize whatever was in the lamp. He also presented another device which did something similar but transferred light through gas between two electrodes. Tesla invented what was essentially a laser. What does laser technology have to do with wireless communication? They both involve transferring energy and potentially information without wires. Tesla claimed that electromagnetic signals could be transferred very long distances simply through the air allowing for global communication. He also proposed that a receiver could be designed to receive electromagnetic radiation emitted from an AC source and convert it into a signal that could be interpreted by a DC system. British scientists were very interested in his ideas and made attempts to replicate his inventions. Tesla stayed behind for a short time to help them make his coils and other inventions which he had made.

One scientist, Sir William Crookes, was particularly interested in Tesla's ideas and allowed him to stay and use his lab for a while before he had to go to Paris to give another lecture that he had promised. The two men quickly became good friends. After a day in the lab, Tesla and Crookes would have long talks on many different topics ranging from wireless communication to spiritualism to Tesla's homeland. During these talks Crookes helped Tesla develop the idea of not only transmitting information across long distances but voice as well. This was the beginning of Tesla's ideas for transmitting radio a few years later. Tesla also revealed his goal to Crookes of eventually controlling weather. They talked about how they could reduce the amount of rain in Britain and increase it over deserts as well as control the distribution of clouds and fog. Crookes also introduced Tesla to the realm of the paranormal. Crookes was a member of the Society of Psychical Research. Crookes claimed to have a good deal of evidence for psychic phenomena including telepathy, séances, and ghosts. Tesla was very bothered by these ideas. He had radical ideas but not that radical. He was still a strict materialist who was made uncomfortable by the spiritual talk which seemed to threaten his worldview in which the universe was simply a machine governed by physical and mathematical laws. Crookes, meanwhile, encouraged him to visit the mountains of his homeland and to reconnect with the spiritual traditions of his ancestors.

In February of 1892, Tesla left for Paris to give a talk before the International Society of Electricians. While there, he discussed alternating currents and emerging medical uses of electricity known as electrotherapy. During this time, he also me the Prince of Belgium who was interested in providing his country with electrical power. Tesla's stay in Paris was interrupted however when he received word that his mother was dying. It was during this time that Tesla had one of his strangest experiences.

As soon as Tesla received word of his mother's condition, he got on a train to Gospic. During the train he did not sleep at all even though he had recently been having sleeping spells due to mental overexertion because of all of his talks and of course overworking for the last seventeen years since the days when he would study nineteen hours a day as a freshman at the Graz Polytechnic back in 1875.

Upon arriving he was greeted by his sisters and their husbands who were all Serbian Orthodox priests. In contrast to Nikola Tesla himself, most of his family was very involved in the religious establishment of eastern Europe and embraced a spiritual worldview. Shortly after he arrived, his mother passed away. It was during this time that he had another episode of the strange illnesses that tended to befall the equally odd Nikola Tesla. He began to suffer from amnesia where he began to forget about most of his life before returning home but he still remembered all of his technical an engineering knowledge. The night before his mother's death, he wasn't feeling well and was carried by some of his relatives to a room where they left him. His thoughts turned to his mother knowing that she would die soon. He reasoned that if she did die, she would provide him a sign. While he was laying down he saw a shimmering cloud cross the room. On the cloud were angels and one of them smiled at him looking upon him with a look of love. This angel began to resemble his mother. During this vision, he also heard the sweet music of angels. The apparition of the cloud and the angels crossed the room and then was gone. What had he just seen? Was Crookes right about the existence of a psychic netherworld? Tesla could not accept this and for the next several months put a lot of thought into searching for a natural explanation for this experience. He had to do this in order to maintain the integrity of his materialistic worldview. He even consulted Crookes for help. He eventually decided the apparition had been a sort of dream he had based on a painting that he had seen. He had been thinking about his mother and so his thoughts about his mother's death and his knowledge of the painting were fused in his mind and in his half-waking state he dreamed up the phantom cloud and the angels. The singing he heard, he reasoned, must have come from hearing the choir of an early morning mass. It seemed reasonable enough to him though I supposed we will never really know what he actually saw.

After his mother's death, it took a few weeks for him to recover in strength to leave his hometown. After recovering, he visited family members and then went to the city of Belgrade, the capital of Serbia at the time to receive a hero's welcome. As a Serbian scientist, he was now a great source of ethnic pride to the Serbs who saw him as a national hero. He also went to Budapest to see old friends whom he had left nine years earlier to go work for Edison Continental in Paris. Upon his return to New York, a triumphant welcome celebration was planned for the returning genius and electric god. Tesla, however keeping to his normal solitary self, simply went to a hotel room in Gerlach and spent the night alone. While traveling in Europe, he had discovered that social interaction had prevented him from being able to make as much scientific progress as he had wanted. He recounts that he felt that he had wasted two years of his life working for Westinghouse between 1889 to 1891. For this reason, he decided to withdraw further into solitude and vowed that he would not work for anyone. He does not appear to have done this out of misanthropic reasons though. He appears to have genuinely believed that isolating himself would be better for the human race since it meant that he could produce more inventions and thus be more useful to humanity as an inventor. This reflects the altruistic nature of Tesla. It does however lead one to wonder if perhaps this was to some degree moral insanity. After all isolation meant that he could not help others in the form of taking on students and passing on his knowledge to younger inventors. In fact, it is arguable that if he hadn't resorted to such isolation that he would have been more useful to humanity in that more of his ideas may have been implemented instead being forgotten after his death and only just rediscovered today. I will leave that discussion, for now, to the philosophers.

Chapter 12

The Electric God

Nikola Tesla continued to pour out inventions in the coming years and continued to grow in fame as an eccentric, reclusive genius. In the Chicago World's Fair in 1893, the alternating polyphase motor which he presented before the Royal Society the previous year was for the first time on public display for crowds of spectators as well as the device that he used to impress his financial backer Charles Peck in which he used magnetic fields to make an egg stand. At the fair, Tesla used his AC polyphase current to pass 1 million volts of electricity through his body without harming himself. The crowd was very impressed. It was also something of poetic justice. Four years earlier, Edison had tried to discredit alternating current by portraying it as dangerous and unwieldy, now Tesla showed that it was so safe that he could pass millions of volts through his body with alternating and not get injured. This was another way in which Tesla could celebrate his victory over his rival, Edison. Furthermore, AC generators based on Tesla's design were being implemented worldwide by electric companies and Edison's DC technology was now on a fast decline. Edison in fact no longer even had control over his own company which now sold AC systems. You could say that the Wizard of Gospic had triumphed over the wizard of Menlo Park. Thomas Edison, despite all of his fame and all of the accolades he had received was ultimately bested by this solitary genius with an odd moustache and crazy look in his eyes from the mountains of Serbia.

The next major accomplishment of Tesla was harnessing the energy of Niagara Falls. Niagara Falls was a project that had been difficult to get off the ground. It began in 1886 when a commission was made to find a way to harness the energy in the enormous waterfall. It was estimated to have the potential to produce 4,000,000 to 9,000,000 horsepower. The project was however rather slow in its implementation. The project was eventually taken over by the Cataract Construction Company owned by Edward Dean Adams which formed the International Niagara Commission. The International Niagara Commission offered a prize of $3,000 to anyone in the world who could come up with a viable way to harness the energy of the waterfall. Lord Kelvin was of course made the chairman being a prominent scientist. Problems were encountered however. When the bid was first made in 1890, none of the major electric companies submitted a proposal. Westinghouse in fact thought that the project was vastly underpriced saying that they were "giving a three-thousand-dollar reward for what was a one hundred-thousand-dollar project." Another problem was that most of the companies were proposing the use of alternating current while Adams preferred direct current. Many proposals were submitted and there was a lot of argument over the best approach to harnessing the power of the waterfall as well as the best way to use it. One of the benefits of the dam was that it could be used to power the city of Buffalo which was only twenty-two miles away and possibly even New York City. If alternating current was used, this was a viable option. But if direct current was used it, would not even be able to reach Buffalo let alone New York City. Eventually the Cataract Construction Company chose to develop a system which would allow hydroelectric power to be gained from the waterfall.

Two companies, the Westinghouse Electric Company and the General Electric Company, which had been the Edison General Electric Company, each proposed a bid for a power system producing up to 15,000 horsepower with three generators. Tesla's polyphase alternating current motors were implemented in the generators and they were successfully used to power Buffalo, New York, and the surrounding area. Similar power systems were built across New York and other places which were used for street lighting, railways, and the subway among other things. Some of them had alternators to convert between AC and DC while some just switched completely to AC. Although Tesla was only an advisor and was not directly involved in the project which was mainly funded and completed by the General Electric Company and the Westinghouse Electric Company, it was his inventions which allowed for it to be a great success. By 1896, Westinghouse was able to add another 50,000 horsepower to the power system. Tesla had succeeded in accomplishing the boast he had made as a boy when he said that one day he would go to America to Niagara Falls and produce power. This seems to have fulfilled a prophecy, albeit a self-fulfilling prophecy. Tesla, during the project, had purposely involved himself and pushed for the use of the AC polyphase generator even granting purchasing rights of the AC generator to General Electric. With Another one of his boyhood dreams accomplished, it was now time to move onto an even more ambitious one, wireless communication and power for everyone.

As the 1890s progressed, Tesla became increasingly preoccupied with the idea of wireless communication. He wanted to create a way to transmit electrical signals without wires. This had long been an aspiration of electrical engineers and electricians, to do away with wires and make electric power something that could be spread across the earth. Alexander Graham Bell who invented the telephone in the 1880s had produced such a device in 1881 which he called a photophone which worked by using sound waves from someone's voice to cause a mica mirror to vibrate and reflect sunlight into a reciever. In 1882, Professor A.E. Dolbear was able to make a wireless receiver where he used coils to produce waves and carbon nodules touching a metal disk to detect the waves. The English scientist William Preece also produced a similar device, but most of the wireless devices produced so far had very short range and were not practical long range communication. Wireless communication at the time was in some ways seen the way that nanotechnology is seen today, something that seemed theoretically possible but that in a practical sense was still a far off dream of futurists. A technology with an enormous potential for application but seemingly insurmountable practical barriers, this definitely sounds like a job for Nikola Tesla and Tesla certainly agreed. He in fact worked feverishly on it not wanting anyone else to come up with the solution before he did. He was justified in thinking that someone might come up with the invention before he did. As he was working on his wireless receiver, Guglielmo Marconi, the Italian inventor who is given much more credit for his work on the radio than Tesla, was working on his own device and created a device very similar to that of Professor Dolbear which he finished around 1896, a year before Tesla made his first long range wireless transmission.

In 1893, he began to experiment with electrical signals. He found that electrical signals have the same properties as musical notes with different frequencies which if he limited the range he could pick up specific frequencies. He began using this property while attempting the transmission of signals over increasingly larger distances. The naturally ambitious Nikola Tesla drew up plans for a transmitter and a receiving station to receive the signal that he intended to send. Before he could send it however, an enormous setback occurred. On March 13, 1895 a fire went through his laboratory and destroyed most of his equipment and much of his previous work, mementos, prizes, articles, notes, and many, many inventions were lost forever. Considering the caliber of Tesla's genius, the destruction of his laboratory was a catastrophe on par with the burning of the Great Library of Alexandria. Much of what he could have given to the world was probably lost at that moment. Not only was this a tragedy for mankind, it was also a personal tragedy for Tesla since it was his life's work. Up until this point Tesla's entire life had been his work. He almost never left the lab and spent almost all hours of the day experimenting and unraveling the mysteries of nature. He had few friends, no romantic relationships, and very few interests outside of his scientific pursuits. The destruction of the laboratory meant the destruction of a part of himself. Tesla was also in a difficult financial situation since he did not have enough income to build a new laboratory to continue his work. The only income he had were occasional royalties from patents in Europe and some funding from benefactors such as the Westinghouse Electric Company. Edward Dean Adams, leader of the commission to gain electric power from Niagara Falls came to his rescue. He offered Tesla a hundred thousand dollars in stock as well as connections with J.P. Morgan a major philanthropist and patron of scientific research. Tesla for mysterious reasons, perhaps because he wanted to remain independent refused this offer but did agree to an offer of $40,000 to fund his research.

It took Tesla two years to rebuild his laboratory and all of the experimental apparatuses which needed for his research. They all needed to be custom made and could not be purchased from any scientific instrument companies. By Spring of 1897, incidentally ten years after overcoming a previous setback in which he had been left penniless by untrustworthy business partners, he had built a new laboratory at Houston Street. In the laboratory he built a transmitter and a receiver. He placed the receiver on a boat which he had sailed down the Hudson River to a distance twenty-five miles from the laboratory. The receiver successfully picked up a signal from the transmitter in the Houston Laboratory. This showed that wireless communication could be achieved over large distances. This was the beginning of modern Radio. In our modern age where wireless internet is commonplace and it is trivially easy for someone in San Diego, California to talk to someone in London, England just through wireless communication assisted by satellites orbiting a thousand miles above the surface of the earth in geosynchronous orbit, it is hard to appreciate just how momentous this event was. Wireless communication had been shown to be possible by inventors in the past such as Guglielmo Marconi, another pioneer in the development of radio communication, but not practically feasible. This discovery would be the equivalent of an inventor building a working warp drive engine today. It showed that a technology which seemed almost fanciful was possible and could be practically implemented. James O'Neil, the first biographer of Nikola Tesla thought that the article in the *Electrical Review* announcing Tesla's accomplishment was quite underwhelming and a bit of an understatement. I would generally agree with this statement. Tesla's plans went beyond the way wireless communication is planned today however. Tesla had an idea that he could transmit electrical signals through the earth itself without the aid of wires and not just through the air. He believed that the earth was like any other body, a charged body with positive and negative charge and that this bipolarity could be exploited to provide power and information to any point on the earth.

Tesla outlined this interesting idea in lectures he gave in 1892 and 1893. Scientists know today that the earth does have electric charge and that electrical energy can be passed through it. It is not however known how it would be done the way that Tesla planned on doing it. He claimed to have drawn up plans for implementing this idea, but these plans have not been recovered.

Tesla, weary that someone might steal his idea and be the first to discover a means of long range wireless communication before him, waited until September 1897 after he had secured his patents before making his inventions fully known to scientists and to the public. He was not usually this paranoid, but it seems to have been especially important to him in this case that he, not someone else was remembered as the one who brought wireless communication to mankind. Once the patents were secure, he explained in detail his invention. What he described are the circuits and transceivers that are still used in radios today. Tesla was not the only scientist to work on radio of course, Marconi also made many accomplishments in the field, but anyone who listens to the radio or uses any form or wireless communication is to some extant indebted to Nikola Tesla.

Other scientists such as the English scientist Oliver Lodge and the Italian inventor Guglielmo Marconi were making similar discoveries at the same time. Both of them relied on transmissions with shorter wavelengths and thus higher frequencies. Tesla reviewed their methods and found that their methods were less effective than his approach. The two scientists eventually had to concede and use Tesla's method. This would bring Tesla into a controversy with yet another inventor and scientist, the aforementioned Guglielmo Marconi, though this time, it would be Tesla who was the distinguished, older invontor whose seniority was being threatened and it would be Marconi who was the young scientist threatening his position over radio patents. This was a battle which would eventually favor

Marconi and indeed it is Marconi's name not Tesla's which is typically associated with radio.

I could write this entire book about Nikola Tesla's inventions and the many ideas he had, but since I have to leave room for another luminary who was a young adult attending college in Zurich at the same time that Tesla was inventing wireless communication, Albert Einstein, I will just focus on a few of his greatest ideas. As time went on the Tesla's proposals became increasingly more extravagant. In additions to sending electric signals through the earth, he believed that, using wireless electricity, the entire atmosphere could be turned into a giant lamp. In 1892, Tesla invented a lamp which operated like the sun. The carbon-button lamp. It was essentially a transparent globe with a lump of carbon or button at its center. The carbon was attached to a wire through which an alternating current was run. This caused the carbon heat up. As it did the gas inside, mostly helium and neon, would begin to absorb the energy of the carbon as the gas molecules bounced off of the carbon mass. This would cause them to speed up and gain energy. Eventually they would begin to glow. The lamp worked the same way as the sun, being illuminated by causing gases in the globe to become hot enough to glow. This is basically the idea behind non-incandescent filament lightbulbs used today. From this phenomenon, Tesla got the idea that the entire atmosphere could be illuminated through the use of electricity. While the lower atmosphere acted as an insulator, the upper atmosphere was fairly conductive, according to Tesla. The solid surface of the earth can also act as a conductor. Tesla believed that a current created between earth's upper atmosphere and the earth's surface could create illumination in the upper atmosphere that would make all other electric lighting obsolete. A way to ionize the atmosphere would simply have to be found to start this current. Tesla, in one letter to an inquirer, suggested using an emitter to shoot X-rays and gamma rays into the upper atmosphere to ionize it and create charge. His exact plans for doing this however have not been recovered.

Chapter 13

The God of Thunder Goes to Colorado Springs
Tesla now knew that it was possible to send wireless signals over large distances for it to be a practical means of transmitting power and communication. He also now knew that it was theoretically possible to attain global wireless communication. He now needed to provide enough power for this communication. He continued to increase the voltage output of his laboratory on Houston Street until he was able to produce millions of volts. He realized however that he would soon need a larger laboratory to conduct such large scale voltages and to produce such enormous power. Additionally, the citizens of New York probably wouldn't be completely comfortable with such high voltage being produced right next door to their homes. His lab experiments were getting increasingly hard to contain with electric sparks and bolts climbing up the walls. By 1899, Tesla was already almost out of the $40,000 that he had received from

Westinghouse back in 1895 and was running out of places to turn. Additionally, he wanted a larger laboratory, preferably one in the country so that he could take advantage of the larger spaces. Help came from Leonard Curtis, owner of the Colorado Springs Electric Company who made him an offer of $30,000 to build a laboratory at Colorado Springs and conduct his research there. This was just what Tesla wanted. In March of 1899, Tesla moved from New York to Colorado Springs and begin building his laboratory. It was much larger than the one in New York City and the subject of much larger scale experiments than his New York laboratory. It is also where he made some crucial discoveries relevant to his plan to ensure wireless global communication. The open space and dry atmosphere of the highlands of Colorado was ideal for his electrical experiments. Colorado is known for its enormous electrical storms. With the help of his assistants, Tesla able to create artificial lightning in an effort to better understand natural lightning.

At one point during his year-long stay in Colorado Springs, in a scene reminiscent of scenes from films about mad scientist, Tesla created enormous coils to produce artificial lightning from a 200-foot mast at his Colorado laboratory. He stood at the edge of the room and told his assistant to throw the switch. "When I give the word, you close the switch for one second and no longer." Finally, Tesla, once he was out of the way of the current, said "Now." The result was what he described as a crackling electrical fire. The next time he did it while standing outside watching the metal ball atop the mast earnestly waiting to see what would happen as the current was activated. An enormous bolt of lightning came from the mast, then another and another followed by crackling thunder. Tesla clenched his fist and gave out a cry of victory. He had done it! Tesla was now a rival of Zeus himself. This spectacle however did not endear the locals to him who in the middle of his experiment cut off his power since he had just nearly destroyed their power station. That is one of the drawbacks of being a thunder god; you tend to be thought of as dangerous.

It was also at this laboratory that Tesla sought to answer the question as to whether the earth conducted electricity. The natural lightning was not always an ally; several times the laboratory was almost destroyed by the lightning storms. This wasn't helped by the fact that Tesla did all the he could to attract lightning. During one lightning storm, Tesla saw a storm approach in a way that he would expect if earth were acting like a conductor. This proved to him that earth was in fact a conductor of electricity. If Earth were not a conductor of electricity it would make it harder to create a current through the earth as he had planned. Though not impossible, it would require much more electrical energy being poured into the earth to charge it before it could be made into a good conductor, sort of the same way you have to transfer a lot of electricity into a foam plate in order to make it charged.

By 1901, Tesla needed a new laboratory to replace the one on Houston Street which he had used for the previous six years. He also needed one in a better position than Colorado Springs. This was the beginning of his Wardenclyffe Tower. Wardenclyffe Tower was a receiving tower from which Tesla intended to transmit signals across the world. It would be a lab that he would continue to use until 1917. He put it into operation and began to transmit and receive signals. There is a strange story that he received an unfamiliar signal. Some hypothesized that it was from Mars. More likely, he may have picked up the messages Marconi was sending around the same time while in doing scientific work in Europe.

After the demolition of his tower in 1917, Tesla continued to be active as an inventor and scientist for the next twenty-six years until his death in 1943. He continued his reclusive habits, spending most of his time in the laboratory and into his old age never pursued romance. Towards the end of his life he did however become emotionally attached to pigeons which he kept as pets. One particular pigeon he found injured and nursed it back to health. After nursing it back to health he gained an odd attachment to it, probably coming from the fact that he had very few human relationships. He once said that he thought of her, it was a female pigeon, as his wife. It appears that he did find a form of companionship, though in a rather strange way.

In 1937, as the situation between Germany and the surrounding European countries began to intensify, Tesla announced plans to build a directed energy weapon. Directed energy weapons, essentially death rays, are one thing that has earned Tesla his mad scientist status in popular culture, that and creating enough volts to take out a national power grid. He claimed that he was going to produce a weapon and give it to every nation and that this weapon would ensure world peace because it would be too dangerous to use. In a limited way, this concept was fulfilled through the invention of the nuclear bomb. Granted, nuclear weapons have not prevented war, but they have prevented wars on the scale of the Second World War. Nikola Tesla, even though he was thought of as weak and sickly and his family constantly worried about his health as a child, he appears to have outlived all of them, dying at the ripe old age of eighty-six in 1943.

To conclude this section, I will discuss how this man's life relates to the life of the man that I covered in the previous part of this book, Sir Isaac Newton. Tesla never worked on gravity or astronomy or astrophysics, unless the story about him supposedly talking to Martians counts, but his materialistic worldview was one which he inherited from Newton. Isaac Newton implied that the entire universe could be understood in terms of mathematical and physical laws and could be understood as if it was an enormous machine. Nikola Tesla essentially took this idea to its logical conclusion.

If the universe was a machine, then like any other machine, it was possible to optimize it for the benefit of humanity. Tesla intended to do just that. If there were no Isaac Newton or Rene Descartes, then there would be no Tesla and no alternating current or widespread use of electricity. It seems at this point that Tesla's work is a vindication of the Newtonian worldview that the universe is a machine and can be treated like one. That is not however the end of the story. Newton was right about the universe being understandable through physics and mathematics, but Newton's view of the universe did not give scientists a complete picture and right around the time that Tesla was making his discoveries, scientists were discovering the deficiencies of Newtonian mechanics. Another genius would have to come along and complete the picture. This brings us to our next subject, Albert Einstein.

Part 3
Albert Einstein

Chapter 14

A Sage is Born: The Young Albert Einstein
Albert Einstein is probably one of the most well-known geniuses of the 20th century. Today his name is almost synonymous with genius and intelligence. His name is also synonymous with physics. What is ironic is that most people who aren't scientists who know Einstein as a genius probably don't know why he is considered a genius. They know that he originated the equation $E = mc^2$ but they are unlikely to know what that equation means. Also Einstein seems to be at the center of numerous misconceptions. Everyone who wants to give their agenda credibility tends to claim that Einstein was one supported their idea. Proponents of the "Law of Attraction" even claimed that he knew about their universal Law of Attraction in which anything you desire will magically become yours

if you want it enough. They don't provide any evidence that he knew about it or believed it of course. There are also those who try to say that Einstein was religious, a creationist, or whatever their belief is that they want to promote. It is understandable why there is so much confusion over Einstein's religious and political views because they are decidedly complex. He was a pacifist for certain and he talked about God but he doesn't appear to have believed in the personal God associated with Judaeo-Christian tradition. His scientific views on the other hand are straightforward and central to what today is called modern physics. It could be said that in the same way that Newton founded classical physics based on his three laws of motion and ideas about force, Einstein is one of the founders of modern physics. His idea of special relativity published in 1905 and general relativity published in 1915 were a solution to the failings in classical models of physics and allowed science to start to move beyond Isaac Newton.

Einstein helped show how the universe is even stranger than Newton himself imagined.

Albert Einstein was born in Ulm in the kingdom of Wurttemberg which was within the old German Empire to Ashkenazi Jewish parents, Hermann Einstein and Pauline Koch. When he was born his grandmother is reported to have thought that his head was "much too fat." Later, after his death his brain was actually removed for study to figure out how he was so intelligent. Shortly after the young Albert was born, his parents moved to Munich in early 1880 where Hermann and his brother Jakob set up an electrical firm. Albert, like the two previous luminaries that we have met, was rather solitary as a child. He tended to keep to himself, usually playing by himself, and from even a young age had an independent spirit. His parents had originally been afraid that he was mentally deficient because there was a delay in his speech. He began to speak around the age of two or three. The only other child he was close to was his sister, Maja. One peculiar aspect of young Albert's behavior was that until about the age of seven he would throw temper tantrums during which his nose would turn a peculiar white color. He would occasionally throw things at his sister and probably other people during these trantrums. Apart from the tantrums, Albert Einstein was not a very disruptive child and was usually very quiet. He attended Catholic elementary school while he was young, it was at the Catholic school that he went through a period of being fascinated with God and religion. Einstein was enamored by the beauty and order which he found in the universe and saw it as reflecting a greater power or intelligence at the time. His parents were proud of their Jewish heritage but not particularly religious and they did not try to control Albert's religious inclinations. Although by the age of twelve, Albert had become a freethinker and religion no longer interested him, he never lost his fascination with the order and harmony that he saw in the cosmos. At the school, Albert was a reasonably good student though slow-working, this is a feature of Einstein's approach to study that would continue later into his life. At school, he kept to himself playing games which required a lot of effort and patience such as building houses of cards. Albert Einstein by all reports had a happy childhood. In Munich, he and his sister lived with their parents in a house with a garden and large trees. His mother was musically inclined and ensured that they both received a musical education. Maja learned the piano and Albert took violin lessons until he was thirteen. Into his later years as a professor at Princeton University, he still was known for playing the violin.

In 1888, he began his secondary education at the Luitpold Gymnasium. While at the Gymnasium, Albert Einstein developed his interest in mathematics, particularly calculus and geometry. He also became interested in physics, an interest for which he would become immortalized. His interests were encouraged by his uncle Jakob with whom he would spend hours talking about science. Albert also read a lot of popular science which further stoked his blossoming interest in science. Although Einstein liked learning, he did not like being at school. The teachers emphasized rote memorization and taught in such a way that Einstein felt like they were trying to stifle his creativity and later he would compare them to drill sergeants. He also made few friends and considered them to be servile to the authoritarian teachers, or at least teachers he considered to be authoritarian. His teachers also had trouble dealing with the young Albert Einstein because he was quite independent by nature and did not like authority. Later in life he became a democrat and a pacifist, this reflects his distaste for authoritarianism and war respectively.

In 1894, Hermann Einstein and his wife and brother were forced to leave Munich because large electrical companies threatened to put them out of business. Hermann Einstein's company still used direct current rather than the more standard and more efficient alternating current. Albert Einstein who was still in school had to stay behind in Munich. Einstein did not enjoy this time being separated from his family, he had the famous Nikola Tesla to thank for this as readers of this book know, and by 1895 he had dropped out and went to Italy to join his family in Pavia. Albert continued his secondary education on his own preferring to learn on his own without relying on authoritarian teachers. In 1895, at the age of sixteen, Albert Einstein took the entrance exam for the Swiss Polytechnic School in Zurich. Einstein did poorly on most of the exam however he did very well in the parts of the exam pertaining to science and mathematics. After this, the school was interested in admitting Albert Einstein but advised him to study secondary school for another year before enrolling. During 1895-1896, Einstein attended the school in the Canton of Aurau, there the atmosphere was less authoritarian and the teachers allowed the students more freedom. The school was also known for having excellent science education. Einstein enjoyed this school much more preferring it to the military like school he had attended his Zurich.

Chapter 15

Age of Miracles

Finally, in 1896, he enrolled in the Swiss Polytechnic School of Zurich as a physics major. Although Albert learned a lot at the school, most of what he learned was self-taught. Albert Einstein tended to skip out on classes and ironically, his fellow students considered him to be lazy and probably didn't think that he would amount to anything. Albert however, was learning a great deal on his own from reading scientific papers on cutting edge theoretical physics. While at the school Albert met Mileva Maric, a female Serbian physics student, and the only woman of Albert's class. She shared his interest in theoretical physics and she would occasionally join him during his self-studying. They soon became close friends and eventually developed a romantic relationship and even had a daughter during the time that they were at the school who either died young or was adopted.

While Einstein enjoyed his time in Zurich, it was not without drama. Einstein's independent spirit caused friction between him and his teachers. It was partly for this reason, that and the fact that he was Jewish that upon graduation in 1900 he did not receive an academic position even though everyone else in his graduating class did. Einstein spent the next few years going through various temporary jobs including one teaching job. In 1902, he was able to get a job at the Zurich patent office as a technical expert. Although this job was far below the skill level of Einstein, the relatively little work required from the job gave Einstein a lot of time to think. It could be argued that this was in fact better for Einstein. If he had been a graduate student under some professor, his independent nature and the tendency of professor's to want to advance their own work and for students to follow in their footsteps would have made life more difficult for Einstein. He would have had to have dealt with a lot of politics and would not have been able to do as much thinking. The patent office allowed him a relatively stress free environment where he could simply think and ponder the mysteries of nature. He also later said that he was able to think more deeply. He believed that at graduate institutions students were pressured to publish a great deal of scientific papers in little time. This, he believed resulted in them producing a lot of papers that were of relatively poor quality and depth. He, at his desk at the patent office, could spend a lot of time to write one paper of the most depth possible which is precisely what he did, well, four papers actually.

In 1903, Einstein married his college sweetheart Mileva Maric. They would remain married for the next eleven years and have two sons, Hans Albert and Eduard.

While at the patent office, Einstein began to ask three questions which he had been asking for the last couple of years. The first one was essentially what was the relationship between light and space? Einstein had always wondered what it would be like to travel on the tip of a light ray. If you were going as fast as the light ray what would you see? The second question he asked was why all objects fell at the same rate? Why did gravity cause the same acceleration in all objects? The third question he asked was whether there was a way to combine gravity and the other forces of nature into one unified force, the subject of the theory of everything.

His answer to the first question eventually led to him to develop the theory of special relativity which he published in 1905. Galileo and Newton had come up with a form of relativity where acceleration depended on frame of reference. A frame of reference in physics is simply the perspective from which you are experiencing motion. If you are on a rocket that is traveling into orbit around the earth, then the rocket is your frame of reference and you interpret all motion relative to the rocket. If you are standing on the surface of the earth watching the rocket, then everything you see is from the perspective of your position on the surface of the earth. From the perspective of the person on the rocket, the rocket is stationary and the surface of the earth is moving away. From the perspective of the person standing on the surface of the earth however, it is the rocket that is moving and the earth that is stationary.

A foundational aspect of physics going back to Galileo and Newton is both of these reference frames are equal. It is possible to do physics from either perspective and get the same answer. By the time of Einstein, physicists had begun to notice that there were deficiencies in the Newtonian views of the universe. According to classical Newtonian mechanics acceleration and velocity are relative to the frame of reference. The speed of light however did not appear to change with the frame of reference. No matter what the frame of reference was, the speed of light was always the same. The problem with this view is that according to Newtonian mechanics the one thing that does not change from frame of reference to frame of reference is time. Time is constant regardless of what frame of reference you use. What if it wasn't however? What if time changes with changing reference frames? This is part of the way in which Einstein revolutionized physics. What he proposed was that the speed of light was constant and that time was different from reference frame to reference frame. He also proposed that the speed of light was in fact the ultimate speed limit, nothing could go faster than the speed of light. Why? Another discovery by Einstein might help answer this question. Einstein, as is well known, also discovered that mass and energy are equivalent. That is the meaning of $E = mc^2$. Since mass and energy are equivalent, mass increases with energy. A cake in an oven actually increases slightly in mass as it heats up. The increase is infinitesimally small but still there. It takes energy to accelerate an object and the higher the velocity the higher the required energy. One more fact I should mention before I tie this all together is that there is a limited amount of energy in the entire universe.

Since it takes energy to make matter move, the amount of energy in the universe places a limit on how fast matter can go. For matter to travel faster than the speed of light, it would take more energy than the energy available in the entire universe. Another reason matter cannot go faster than light is more theoretical. It is that as one approaches the speed of light, time slows down. Theoretically, if something or someone were to attain the speed of light, time for that object or person would stop. Since you need the passage of time in order for there to be speed, traveling at the speed of light becomes a logical as well as theoretical impossibility.

If one accepts this weird conclusion that time slows down as an object goes faster, then physics only get weirder from there. Einstein talked about the twin paradox which has been elaborated since Einstein proposed it. Suppose there are two twins. One becomes an astronaut and goes on a voyage to a distant star system at a speed close the speed of light while another stays on Earth. The twins are twenty years of age when the astronaut twin leaves. He returns fifty years later when both of them are seventy. According to the theory of special relativity the one who stayed on earth would be an old man but the one went to space for fifty years would barely have aged at all. The reason is because the one going at a speed close to the speed of light would have experienced less time than the one on Earth.

Another strange fact about speeds very close to the speed of light is that colors change as you go fast. Light, as Newton discovered is made up of a spectrum of light of different wavelengths. One thing that Newton did not discover however was that different wavelengths are associated with different energy. High energy light has shorter wavelengths while lower energy light has longer wavelengths and depending on whether you are come towards an object or going away from an object, the wavelength will be different. If you are moving away from an object at a speed close to the speed of light, the light behind you will have longer wavelengths than the light in front of you. As a result, items behind you will look more red, since red light has a longer wavelength and light in front of you will look blue because it has a shorter wavelength. This is called the Doppler shift. To use a common analogy, let us suppose that you are by a railroad track and a train comes passes you. As the train comes near you, you will notice that it has a higher pitch than when the train is behind you. The reason for this is that the sound waves in front of the train are being pushed closer together because of the pressure of the oncoming train through the air. The sound waves behind the train on the other hand are farther apart because there is less pressure to push them together. The process for light waves is very similar. As an object moves at a speed close to light, wavelength of the light in front of it contracts becoming shorter and the wavelength of the light behind it increases becoming longer. This is why in cosmology astronomers say that the galaxies are "red shifted." The galaxies are moving away from each-other very rapidly and as a result the light from distant galaxies has longer or redder wavelengths. The light is shifted to the red end of the spectrum. Using our imagination, let us suppose that you were in a car traveling at normal speed and there was a car in front of you and a car behind you and both of them were green. Now let's suppose that for some strange reason you all start accelerated towards light speed. As you got closer to the speed of light you would notice something very odd, as you all approached the speed of light, the car in front of you would start to turn blue and the car behind you would start to turn yellow. This is because yellow light has a longer wavelength than green light and blue light has a shorter wavelength than green light. If you started to accelerate faster so that you were also speeding way from the car behind you and towards the car in front of you, you would notice the car behind you turning red and the car in front of you turning violet as you increased in acceleration. This is what Einstein found.

In addition to this just being very counterintuitive it may hold the secret to interstellar travel one day. One of the major barriers to reaching other stars is the time required. Even at the speed of light it would take 100,000 years to cross the milky way galaxy. A space traveler traveling at the speed of light however would not experience the passage of 100,000 years. The entire visible universe in fact could be traversed in about 50 years' time from the perspective of our space traveler. When the space traveler finally slowed down however, tens of billions of years would have passed and the sun would be no more and our galaxy would have merged with other galaxies. He would have traversed the entire observable universe though. We don't know if Einstein thought much of space travel but his ideas definitely did advance the possibility of future long distance space travel both with his theory of special relativity and as we shall see his theory of general relativity.

There is also another phenomenon on which Einstein gave a lot of thought which, pardon even more digressions into physics, I am aware that you purchased this book to learn about great scientists and not to study for your upcoming physics quiz, but it would be a disservice to the history of science to leave this one out, especially since it is this discovery what got him the Nobel Prize in 1921, the photoelectric effect. During the 19th century it was found by scientists such as Heinrich Hertz that when a metal surface was exposed to light or electromagnetic radiation such as ultraviolet rays that electrons were emitted by the metal surface. Essentially, shining light on a metal produced electrons. When physicists began to study the photoelectric effect they came to some unsettling conclusions because these conclusions contradicted classical Newtonian interpretations of the phenomenon. Basically classical or Newtonian physics would have predicted that the velocity of the electrons would increase with the intensity of the radiation and that beneath a certain intensity that there would be no electrons emitted because there would not be enough energy. What they found, however, was the velocity was stayed the same with regardless of the intensity of the light and that changing the intensity only changed the number of electrons being emitted. They also discovered even when the intensity of the light was so low that there shouldn't be enough to emit even one electron, electrons were still emitted. Equally puzzling, they realized that light below a certain frequency, regardless of the intensity, did not cause the emission of any electrons. This all contradicted classical physics which treated electrons like any solid object governed by Newton's three laws. How were they to explain all of this? Einstein who read many scientific journals and stayed caught up with scientific research pondered these in light of his other musings about physics regarding the speed of light. He eventually came to the conclusion that light was quantized, that is that light, rather than just being a continuous substance like a wave, came in particle form as packets of energy which eventually became known as photons. A photon is basically a particle of light. This explained the strange facts of the photoelectric effect. The reason that the electrons only increase in number with increase in intensity is that more intensity means more photons. More photons allow for more opportunities for photons to interact with electrons and cause them to be liberated from their bonds within the metal. This also explained why intensity of less energy to liberate one electron still produced electrons. Since we are talking about light particles with dimensions similar to the electrons, it is conceivable that a handful of photons could liberate a handful of electrons individually even though it wouldn't be enough to completely ionize the material. This discovery of the photoelectric effect by Einstein along with discoveries and contributions by other scientists such as Max Planck, James Clark Maxwell, Heinrich Hertz among others, helped to change how we look at light and laid the groundwork for quantum mechanics.

Einstein published his ideas about relativity and the photoelectric effect in 1905 while still a lowly technical expert at the patent office in Bern Switzerland. He published them in a series of four articles in *Annalen Der Physik,* a major European German-language scientific journal, that later became known as the *Annus Mirabilis* papers. *Annus Mirabilis* means "miracle year" and the title is definitely appropriate considering that these four articles laid down the foundations for modern physics and all future work in quantum mechanics and relativity. He immediately gained the attention of the scientific community. Some physicists liked his ideas and some did not. The Polytechnic School in Zurich liked it enough that they awarded him an honorary Ph.D. in 1906, Einstein was finally a doctor. Einstein was finally awarded a teaching position as an assistant professor at University of Zurich in 1909. By 1911, he became a full professor at the University of Prague before returning to Zurich the following year. He continued to gain recognition within the scientific community for his scientific work on both relativity and the photoelectric effect. By this time, he had begun working on the next major contribution to his ideas.

Yes, I am going to tell you more about physics, this is about a physicist. I will try to make it brief and interesting. Also, admit it, secretly you really do want to know the details of Einstein's ideas about physics and have them explained to you in a simple comprehensible way. Back in 1907, while working at the patent office, Albert Einstien had realized that he needed to make additions to his theory of special relativity because of gravity. The main issue he had with gravity was the observation made by Galileo in his famous thought experiment of dropping a small object and a large object from the Tower of Pisa, that all objects regardless of their mass have the same acceleration due to gravity. Whether an object is a boulder or a pinhead, in the absence of forces other than gravity, the acceleration is always the same 9.8 meters per second squared or about 32 feet per second squared. This troubled Einstein since it meant that that a reference frame in which there was acceleration due to gravity and one in which there was acceleration that was not gravity but numerically the same as gravity looked exactly the same. Over the following eight years he came to the theory of general relativity. Einstein proposed that gravity rather than being a force was a consequence of the structure of space and time. According to the theory of general relativity, space and time are woven into one fabric naturally enough called space-time. Space time becomes curved or bent by massive objects which warp the fabric of space around them. This bending of space-time near massive objects causes objects to fall towards the object bending due to space time. Consider, as an analogy, a mattress with a bowling ball placed on it. As the bowling ball sits on the mattress, the mattress is bend by the mass of the bowling ball. If you were then to roll a golf ball beside the bowling ball, it would follow the curve of the mattress and roll towards the bowling ball. This is essentially how the gravity between two objects such as earth and the sun works in general relativity.

Einstein was essentially saying that the acceleration due to gravity was not due to any force of nature but because mass somehow deformed space around an object to cause acceleration. This is often conceptualized in 2-D examples such as the mattress and the bowling ball but I probably do not need to explain that it is only analogy, space-time is not two dimensional and the stars are obviously not sitting on a cosmic mattress. Space is deformed in three dimensions. To make matter stranger, scientists have since learned that in the same way that the surface of a globe is a two dimensional surface wrapped around a three dimensional object, the universe appears to be a three-dimensional universe somehow "wrapped around" a 4-dimensional object. I put "wrapped around" in parentheses because it is not wrapped around in the same way as a globe. The bending of space-time implies that not only is space bent but that time is bent as well. This is why that clocks in Earth's orbit actually are a few miliseconds faster than those on Earth's surface. The force of Earth's gravity is slightly stronger at the surface than in orbit. This shows that time actually slows down as gravity gets stronger. The potential for space-time to be bent or distorted by mass led scientists to one of the strangest consequences of general relativity theory, black holes.

Einstein predicted the existence of black holes in 1916 as a consequence of his theory. He conceptualized them as objects that deform space-time to such a degree that the space-time fabric around them becomes like a very deep hole, you could say a bottomless pit, from which nothing, not even light, can escape. The phenomenon in nature that is closest to Einstein's predicted object is a type of collapsed star. Most stars do not become black holes. The smallest stars will become white dwarfs which are basically glowing balls of gaseous iron which were once the cores of stars whose outer layers were blown away by their own solar wind. This will be the eventual fate of our own sun. Larger stars, however, at the end of their lives go out in enormous explosions that for a short time make them brighter than any other star in their galaxy. These explosions are called supernovae (singular: supernova). If a supernova were to occur within the vicinity of our own star, earth would be sterilized by the radiation and life would be wiped out. Let's hope that doesn't happen soon. Many of these stars become what are called neutron stars, stars so dense that the electrons and protons in their atoms have been fused into neutrons. These bizarre objects are the size of cities, about fifteen kilometers, more massive than the earth, and made up entirely of neutrons. A very small percentage of stars which are above a certain size will however experience the most dramatic and bizarre fate of all. After they go supernova, they become objects so dense that they have no size. They are singularities, or basically points in space. They also have such strong gravity that anything that falls within a range called the event horizon can never escape its gravity.

Since we are talking about Einstein, I might as well explore one more of his physics ideas. As often happens in physics, one weird idea leads often leads to yet a stranger idea. If black holes are not weird enough, another implication that Einstein realized later in life was the concept of an Einstein-Rosen bridge, or wormhole. The concept was proposed in 1935. By then, he was living America teaching at Princeton University at the institute for Advanced Studies. A wormhole is basically a tunnel through space-time allowing one to go from one point in the universe to another. Essentially the bottomless pit in the previous example becomes a funnel which reopens carried anything falling into one end of the funnel to a different point. Einstein and another scientist, Nathan Rosen, hence the name Einstein-Rosen bridge, attempted to formulate the idea of a wormhole to get rid of the singularity which was created by the black hole. Einstein as we shall see later, did not like things that were mathematically messy and singularities are quite messy mathematically. The wormhole thought of by Einstein, is very short-lived and would be too unstable for anything to pass through but that doesn't mean that someone won't figure out how to make a workable wormhole someday.

Chapter 16

The Materialist Mystic: Einstein and the Search for Beauty in the Cosmos

Einstein was finally in academia and teaching now, but all was not well. In 1913, Einstein was invited by Max Planck and Walter Nernst to do research at the Berlin Academy of Sciences. Albert Einstein agreed and went to Berlin but his wife Mileva Maric did not want to go to Berlin became increasingly unhappy during their stay. Their marriage had already been strained by the relationship that Einstein had with his cousin Elsa Lowenthal with whom he became increasingly close. The couple separated not long after Einstein's arrival in Berlin. Mileva went back to Zurich with their two sons and Albert stayed in Berlin more devoted to his work than to his marriage. He however did agree to support his wife and children financially by sending them 5600 riechsmarks a year, a good percentage of his yearly income. Meanwhile he became increasingly closer to his cousin Elsa whom he would marry in 1919 after his divorced his first wife.

Not only did his wife leave him in 1914, but a major world event suddenly made his life at lot less comfortable and would thrust him into a topic he had up until then ignored, politics. On July 28, 1914 the first world war began between the central powers of the German Empire, Austria-Hungary, and the Ottoman Empire and the allied powers of Britain, France, and Italy. Germany at this time was a world power with colonies overseas, a powerful military presence across Europe, and a thriving economy. Germans were patriotic and took pride in their military superiority. The militarism of the German Empire had always repulsed Einstein, who had gained Swiss citizenship in order to avoid compulsory military service. He had up until this point not gotten involved in politics, but during war the everyone around him exhibited a sense of patriotism and nationalism for Germany. I suppose it could be compared to the atmosphere in the United States right after 9/11. There was a strong sense of patriotism and anyone who didn't share in the patriotism was viewed with suspicion. Einstein took the position of a pacifist which did not endear him to many of his German colleagues who thought that he was being unpatriotic. After the war, Einstein applauded the downfall of the Kaiser and looked forward to the Weimar Republic, hoping that through the Republic Germany would become more peaceful and less nationalistic. He also encouraged young men to refuse military service which did not endear Professor Einstein to patriotic Germans who were angry over having lost the war and having their country humiliated by the impositions placed on Germany by the allies. This anti-nationalist perspective combined with Einstein's Jewish heritage set him on a collision course with a movement which would soon become a powerful force in German politics, the National Socialist Party, also known as the Nazi Party.

Despite accusations of being unpatriotic, Einstein's scientific career continued to advance and he continued to come up with ideas which promised to revolutionize physics. After 1915, his mind turned to his third major question, could gravity and the electromagnetic force be combined in to one overarching force? This question was one which he would think about for the rest of his life and he would never be able to answer. In 1919, his theory of relativity was confirmed when during an eclipse stars were seen which were not supposed to be visible since they would have been behind the moon from the perspective of the earth. The light from the stars had apparently bent around the moon to reach the earth. After receiving the Nobel Prize for his work on the photoelectric effect, Einstein became world-famous. He soon was a household name throughout the world. As the 1920s went on he thought more and more about this idea of a grand unified theory.

Einstein had always been enamored by the beauty and harmony that he saw in the universe. One reason he wanted to find one force governing all of physics was because he thought it was beautiful. Just as Isaac Newton had been able to unify all of physics, celestial and earthly, in his day under three laws of motion and universal gravitation, Einstein wanted to unify the disparate parts of physics in his day into one coherent whole. The electromagnetic force and the gravitational force work in very similar ways. Both are forces governed by an inverse proportion law, though the electromagnetic for can be described by different laws of proportion depending on whether you are talking about a point charge, a sphere, a ring or an infinite sheet. Also the equations describing the electrostatic force between two objects is very similar to the equation describing the gravitational force between them. One significant difference between the two forces though is that while the electromagnetic for can be both attractive and repulsive, gravity is a merely attractive force.

Another important distance is that the electromagnetic force is a much stronger force than gravity by a factor of about a billion. Einstein nonetheless wanted to find a way to combine the two seemingly unrelated forces into one over-arching theory and one set of equations. His efforts became more difficult in that it was eventually discovered that there were more than just two forces, there was also the strong nuclear force and the weak force. This didn't completely dash his hopes of unifying all of the forces of nature into a single force though and to this day there are some physicists who are looking into a way to combine the four forces of nature. Success has in fact been made with the electromagnetic force and the weak force as physicists have found that the weak force is simply another aspect of the electromagnetic force and they have combined them into the "electroweak" force. Another complication that inhibited his efforts was the fact that, according to general relativity, gravity was due to the structure of space-time and not an actual force.

This meant that there would need to be a lot of rethinking of electromagnetism in order to unify it with gravity.

Although Einstein did not find a way to unify the forces, he did not stop trying and he encouraged others after him to keep trying. He was convinced that there was a way to do so because he believed that would make the universe most beautiful and harmonious. The religious views of Albert Einstein are debated even today. Most of the evidence suggests that he was an agnostic or an atheist and skeptical of the idea of a personal God. At the same time, he did have a sense of spirituality and one time said that "religion without science is blind, science without religion is lame." This not say that Einstein was religious but he did seem to have an almost religious appreciation for nature and the laws of nature that are revealed by science. He once said that he believed in "Spinoza's god." Spinoza was what you might call a naturalistic pantheist. He believed that God was identical with the universe and that God was more of a metaphor for the harmony of the laws governing the universe than an actual personal being. If Einstein believed in any deity, it was this one. This does make him like Newton in that sense. Both men ultimately believed that their scientific endeavors were a search for God, the deities that they were searching for were however very dissimilar. Einstein's God was Spinoza's God, impersonal and abstract, while Newton's God was the personal God of Christianity, Judaism, and Hermeticism.

This search for Spinoza's God, meaning the search for beauty and harmony in the mathematical description of the universe, may have actually inhibited Einstein's scientific endeavors. Although Einstein played a major role in developing quantum mechanics as a field, he ended up rejecting it because one of the main principles of quantum mechanics is that exact measurements are not possible and that every measurement is probabilistic to some degree.

Although it was Max Planck who, with his idea of a black body radiation, first suggested that matter emitted energy in discrete quantities which he referred to as "quanta," it was Einstein who took it a step further and said that light was quantized and carried in the form of photons. The idea of photons allowed Einstein to explain the photoelectric effect. Another contribution of his to quantum mechanics was the idea of Einstein coefficients for spontaneous radiation in 1917. Einstein also contributed to statistics in in 1925 with the publication of the Bose-Einstein Statistics. By the mid-1920s however, he realized that the probabilistic nature of quantum mechanics clashed with his philosophy of science. To Einstein, the universe needed to be comprehensible and everything needed to be reduced to a few equations which could predict the behavior of the universe exactly. That was beauty to Einstein. Quantum mechanics which says that all motion on the sub-atomic level is probabilistic and cannot be predicted with certainty would shatter any hope of attaining this view. Though Einstein helped to bring quantum mechanics into being, he never accepted it. Einstein tried to find ways to show that this idea was incorrect. In December 1926, Einstein wrote a letter to the physicist Max Born in which he made his famous quote "God does not play dice with the universe" essentially saying that he disagreed with the probabilistic nature of quantum mechanical predictions about the world. In early 1927 at the Fifth Solvay Conference in Brussels, Einstein participated in a debate with Neils Bohr. Also present at the debate were Max Born and Werner Heisenberg who had just come up with the Heisenberg Uncertainty Principle which said that one could not know the position of a particle without losing information about its momentum and vice verca. This is remembered as a historic debate in which Einstein defended the deterministic view that everything could be understood and predicted with certainty with physical and mathematical laws and Niels Bohr was the champion of the quantum mechanical model that said that predictions could only be known to approximate certainty. What is interesting about this debate is that Einstien, a man who had overturned physics twenty years earlier was now resisting the next major change in physics. One the other hand it has also been noted that his contributions to physics were conservative from the beginning. Although he said that there were certain things which could not be explained by Newtonian mechanics such as electromagnetism and objects traveling at the speed of light he did believe that Newtonian mechanics was approximately correct at low velocity and large scales. Essentially anything that travels much slower than the speed of light and is larger than an atom behaves in a way more or less predicted by Newton's laws. Considering Einstein's view of the universe, it makes sense that he wanted to preserve Newtonian mechanics as much as possible since it did imply a sense of harmony and order as well as determinism in the universe which all appealed to Einstein.

Einstein in his efforts to refute quantum mechanics did a thought experiment where there was a box with a single slit covered by a shutter and clock inside the box. When the shutter opened, the photons would escape. Einstein said that using the clock and the mass-energy equivalence equation, the energy of the photon before and after leaving the box could be measured exactly. He presented this idea during a debate with the Danish physicist Niels Bohr. Bohr at first thought it was compelling but then realized that Einstein had not taken into account the relativistic effects of gravity and how they would affect the accuracy of the clock still making the timing uncertain. Eventually, Einstein gave up on trying to disprove it and simply said that the theory was incomplete and that there had to be more pieces of the puzzle waiting to be discovered.

Toward the 1930s, things began to heat up politically. The Weimer Republic, the successor government to the German Empire after

Germany's defeat in World War II, was languishing economically due to the harsh postwar stipulations placed on the Germany for its former militarism. Sky-high inflation and unemployment caused a great deal of malaise for everyday Germans and many Germans began to become very resentful to the allies for their harsh measures and to the Weimer Republic for its failure to give them a stable lifestyle in which they could support their families. This soon led to the increasing popularity of extremists on both the left and the right. From the political Right arose the Nazi Party which promised to restore glory of the German Empire and exact vengeance on those who had taken away their prosperity. Einstein, a staunch pacifist and supporter of the Weimer Republic was also upset over the treatment of Germany by the allies and was a part of several German humanitarian organizations which called for better treatment of Germany. He opposed the Nazi party and urged the Germans to seek peace and reconciliation with their neighbors. Things got worse as the Nazi party began to target various groups as scapegoats for the German people to vent their anger. One of these scapegoats happened to be German Jews. Einstein's Jewishness made things increasingly uncomfortable for him. He began to speak out against the increasingly anti-Semitic atmosphere which did not endear him to many Germans, and especially not to the Nazis. In 1933, the same year that Adolph Hitler came to power and made himself Chancellor of Germany, Albert Einstein wisely immigrated to the United States.

Chapter 17

Sage of Princeton: Einstein comes to America Albert Einstein first visited America in April 1921. He arrived in New York City where he was greeted by Mayor John Francis Hylan. Over the next couple of weeks, he spent time giving lectures at Princeton and Columbia University. He later said that he appreciated what he considered the "internationalist psyche" of the United States. He made a second visit in 1930 and came again to work at the California Institute of Technology as a research fellow in 1931, 1932, and 1933 for 1 term each year. After the end of the winter term in 1933, Albert appears to have returned to the Berlin Academy of Sciences only find that he had lost his position because of a new law passed by the Nazi party banning Jewish intellectuals from holding any academic positions at German universities. Thousands of Jewish scientists were removed from their teaching positions and many of them were forced to flee the country. Albert Einstein fared no better and his works were also the subject of book-burning. It also around this time the Nazi propaganda minister Joseph Goebbels said that "Jewish intellectualism is dead." There was particular hostility towards Albert Einstein, probably because of his pacifist and internationalist views which did not square well with the national socialist ideology which had taken Einstein's homeland by storm. It is reported in one biography of Einstein that a German newspaper offered $5,000 reward for delivering Einstein to the German authorities to be executed by hanging.

Einstein left Germany without knowing where he would work or where he would live. He just knew that he needed to get away from Germany and the Nazi Party. While staying temporarily in Belgium, Einstein received word from a British naval officer whom he had befriended, Oliver Locker-Lampson, Locker-Lampson welcomed Einstein to stay temporarily at his cottage outside of London where he placed an unofficial guards for Einstein's protection. During this time, Einstein met with Winston Churchill where he urged Churchill to rescue more Jewish scientists from Germany. Churchill responded to Einstein's plea for his countrymen by sending a scientist friend of his to Germany to recruit German Jewish academics to teach at British universities. Over the next year Einstein received offers from British universities and even an offer from the Turkish prime minister to teach in Turkey. The British attempted to offer him a faculty position and Locker-Lampson attempted to advocate to have a bill passed that would provide Einstein British citizenship to become a faculty member at Christ's College at Cambridge. The attempt was however unsuccessful.

In late October 1933, Einstein returned to the United States to take a position as a resident scholar at the Institute for Advanced Studies at Princeton University. In 1935, he chose to become a permanent resident of the United States and applied for U.S. citizenship. He was now a permanent scholar at the Institute for Advanced Studies and would be for the rest of his life. It is during this part of his life that Einstein began to resemble the wizened old genius with crazy hair and casual clothing that appears so much in popular culture. While at Institute for Advanced Studies, he worked with such luminaries as Nathan Rosen and Boris Podolsky. He also became increasingly embroiled in both American and global politics. His stance as a pacifist and a Zionist put him at odds with both the more military minded Americans one on hand and the isolationist America Firsters who also were hostile towards him because of his Jewish heritage. It is notable that at the time that Einstein came to the United States there were almost no Jews employed at major American universities. This reflects the hostility that many in mainstream American society had towards Jews at the time. In light of what had happened to many of his fellow German Jews, Albert Einstein felt it was his duty to speak out against anti-Semitism. He worked with Zionists such Rabbi Stephen Wise to advocate the creation of a Jewish homeland as well as to fight fascism. He believed that it was imperative for the survival of the Jewish people that they be given a homeland where they would be able to preserve their culture and be safe from persecution. This was not a case of Jewish nationalism on Einstein's part however. Einstein hoped that the Jewish community would become a model for the rest of the world to follow in terms of solidarity. Einstein was a staunch internationalist and what he hoped for was a world community lacking aggression. He however believed that to help maintain peace for the Jews, they did need a homeland just like everyone else.

It is during this time that Einstein came closest to becoming the wise-cracking crazy-haired old sage in popular culture. Today Einstein is often thought of as being an aloof isolated genius, not too different from two other personalities that I have outlined in this book. He was like them and always had independent streak but he was not completely isolated and neither was he aloof. Albert Einstein was pragmatic and thought of everything in concrete physical terms before thinking of it in abstract mathematical terms. One reason that he liked the photon in his theory of light was because it suggested a concrete reality behind the math even if this concrete reality made the math a little harder.

He was not isolated either. In his scientific and political endeavors, he would collaborate with like-minded individuals such as the Zionists Rabbi Stephen Wise and Chaim Weizemann and pacifists and internationalists such as Bertrand Russel and Emery Reeves. Collaboration in his scientific endeavors can be seen from many of the concepts which bear his name such as Bose-Einstein statistics and the Einstein-Rosen bridge. Einstein shared credit for both of these ideas with another scientist with whom he collaborated. In this way Einstein was probably the most social of the thinkers that I have explored in this book. This is not to say that Einstein was especially social. he wasn't when compared to most people, but compared to Nikola Tesla and Isaac Newton he had more of social life and appears have had many more personal friends in many countries than Tesla or Newton. Another notable difference so far is that Einstein did not mind sharing his ideas and sharing credit for his ideas with other scientists. This sets him apart from Newton who was constantly paranoid about someone stealing his ideas. There is no Newton-Halley equation or Leibniz-Newton principle. The ghost of Sir Isaac Newton is probably writhing in anguish as I write the latter combination. This also sets Einstein apart from Tesla to some degree since although Tesla was less paranoid about his ideas being stolen, he was very much concerned with preserving his reputation for being the originator of his inventions, but this is a digression.

Einstein did engage in scientific developments during this time. He continued to work on finding a way to reconcile quantum mechanics and general relativity in a way that would allow him to express the behavior of the entire universe as a single set of equations. He also in working with other scientists came interesting conclusions about the implications of relativity. He and Nathan Rosen published a paper on wormholes in 1935 and in 1939 Einstein and several other scientists discussed the properties of black holes in more detail. The 1930s were however a politically tumultuous era and Einstein became increasingly involved in politics as the decade continued. After moving to the United States and seeing the malaise of his fellow European Jews, he criticized the British for curtailing Jewish immigration into Palestine. Things did not look too good in Germany at this point. Under Hitler, Germany was continuing to grow in power and influence. With intention of gaining regaining control of all land inhabited by ethnic Germans and all land ruled by the German Empire before the First World War, they annexed Austria in 1938. The next year in

September 1939, the Second World War started as Germany annexed Poland.

At first Einstein had advocated disarmament and peace, but when he learned that the Nazis were attempting to build an atomic bomb, he decided that if such a weapon was to be built it would better that the Allies had the bomb than the Axis. In 1939, shortly after the start of the war, Albert Einstein and several other scientists including the physicist Leo Szilard wrote a letter to President Franklin Delano Roosevelt warning him that the Nazis intended to develop the bomb. This prompted Roosevelt to begin the Manhattan project which would eventually build the atomic bomb, two of which would be dropped on The Japanese cities of Hiroshima and Nagasaki. Later Albert Einstein would tell Linus Pauling, a friend as well as fellow scientist and pacifist, that this was the greatest mistake of his life though he said afterwards, "but there was some justification, the danger that the Germans would make them."

Although Albert Einstein did not directly participate in the Manhattan project he did gain the title "grandfather of the bomb," even though he later denounced use of the bomb. What is especially ironic is that Einstein, a pacifist, is not only one of the scientists who wrote the letter that prompted the Manhattan Project but also it was in a way his discovery of the equation $E = mc^2$ which allowed there to be a bomb in the first place. The mass-energy equivalence formula essentially says that the energy in an object is equal to the its mass times the speed of light squared. This means that for a little bit of mass, you get a tremendous amount of energy. It is this physical principle that the bomb exploits. Within the bomb, the plutonium goes through a runaway chain reaction that results in an explosion as well as enormous amounts of radiation. It is this process that makes nuclear weapons so destructive. Although scientists probably could have come up with the bomb without Einstein providing the theoretical groundwork, Einstein's theoretical framework did certainly accelerate the development of the bomb. In the same way that Einstein's view of how the universe should be was undermined by his own discoveries in physics, one of his dearest goals with regards to society, world peace, was also undermined by another one of his scientific discoveries. This is perhaps the only part about Einstein's story that is kind of sad. On one hand he was never able to see his vision of the world achieved in the case of pacifism and in the case of relativity and quantum mechanics, he also was not willing to give up on an idea even though it was clearly wrong.

Chapter 18

Final Years of the Sage and Epilogue

By the second World War, Einstein was at the height of his fame. He was well known by all physicists and much of the public. He was scientific advisor to many political figures such as Franklin Delano Roosevelt. It was this fame that allowed him to make many of his political and social views known and promote them. I have already discussed some of his political and social ideas, pacifism, internationalism, and Zionism. As this part of the book as progressed, I have steadily talked less about his scientific contributions and more of his political contributions. This is partly because, as with any scientist, his scientific productivity waned over time. His most creative scientific years were between 1900 and 1919, though you could extend it to 1925, afterwhich he began to slow down. His primary pursuit in the latter part of his life was his search for what he called the theory of everything to unify all branches of physics with ones set of equations. Although he never found it, some physicists are still looking for it. Another reason I am also talking more about his political and scientific views is because they actually did occupy him more during his later years as he became more active in politics. His reason for becoming more active in politics was because the world seemed to be deteriorating before his eyes and as an intellectual and scientist, he felt compelled to do something about it. In 1925 he had helped fund the establishment of the Hebrew University in Jersualem to help Zionist causes and he was a part of the NAACP in Princeton. As you can probably guess this earned him popularity among some but also hostility from others. It didn't help that Einstein seemed to from the perspective of Anti-Semites seemed to confirm a lot of their beliefs about Jews. He was an internationalist, a democrat, a socialist, an anti-racist, a pacifist, and a Zionist. It is a good thing that he wasn't also a banker, or there would probably be anti-Semitic conspiracy theories circulating to this day about him. Well actually there might be. I had better not give them ideas.

At the end of World War II, aghast at what the atomic bomb had accomplished at Hiroshima and Nagasaki, Albert Einstein advocated nuclear disarmament and worked with fellow pacifist and internationalist Emery Reeve to advocate the establishment of a global government to ensure peace between nations and nuclear disarmament. Einstein supported the establishment of the United Nations which were established in 1947. He also became increasingly critical of capitalism and by the 1950s, he considered himself a socialist. His scientific ideas gained him much admiration among everyone with the exception of the Nazis who tried to discredit relativity as "Jewish physics," but his political ideas were quite controversial. His stance as a socialist caused many during the McCarthy era to suspect him of being "Red." He however did not support Soviet Communism and instead supported a form of international socialism which respected individual rights. He did see the establishment of the state of Israel in 1948 and was offered the presidency in 1952 though he turned it down. For the rest of his life he remained a controversial but beloved figure in both within and outside the scientific community. When he was dying due to internal bleeding from an abdominal aneurism in 1955, he asked them to let him die with dignity and not attempt to prolong his life artificially through surgery. He is reported to have said "I want to go when I want. It is tasteless to prolong life artificially. I have done my share, it is time to go. I will do it elegantly." Albert Einstein died on April 18, 1955 in the Princeton Hospital in New Jersey. He was seventy-six, though in those seventy-six years he had revolutionized physics and had left a lasting impression on human civilization which will by no means fade soon.

There you have it, the lives and ideas of Isaac Newton, Nikola Tesla, and Albert Einstein. They were all geniuses, they all were in some way isolated because of their tendency to pursue science above other endeavors, and they all had profound effects on science and society. These three men represent turning points in the scientific age. Isaac Newton represents the moment where modern science began take off in earnest with his theories of motion and gravitation which made not just earth but the entire universe comprehensible as a cosmic machine. Nikola Tesla represents the ultimate consequences of Newton's accomplishments, through exploiting Newtonian mechanics to invent technology that would revolutionize society. Einstein represents the end of this era and yet another turning point in the history of science where the scientific revolution accelerated even farther than it had with Newton. The question to ask now is who will be the next great mind? Who will be the next Newton or Tesla or Einstein? We may soon find out, but for now let us follow in the footsteps of these men and do our best to advance science and society for the better.

Bibliography

Bardi, Jason Socrates. The calculus wars: Newton, Leibniz, and the greatest mathematical clash of all time. Perseus Books Group, 2009.

Calaprice, Alice, ed. The expanded quotable Einstein. Princeton: Princeton University Press, 2000.

Carlson, W. Bernard. Tesla: Inventor of the electrical age. Princeton University Press, 2013.

Darrigol, Olivier. A history of optics from Greek antiquity to the nineteenth century. Oxford University Press, 2012.

Fölsing, Albrecht, and John Stachel. "Albert Einstein: a biography." Physics Today 51 (1998): 55.

R. W. Fuller and J. A. Wheeler, "Causality and Multiply-Connected Space-Time," Phys.

Rev. 128, 919 (1962)

Frank, Philipp, George Rosen, and Shuichi Kusaka. Einstein: His life and times. Da Capo Press, 2002.

Gallos, Joan V., and Walter Isaacson. "Einstein: His Life and Universe." (2008): 594-600.

Highfield, Roger, and Paul Carter. The private lives of Albert Einstein. Macmillan, 1994.

Hoffmann, Banesh, and Helen Dukas. "Albert Einstein, creator and rebel." (1973).

M. S. Morris et al., "Wormholes, Time Machines, and the Weak Energy Condition," Phys. Rev. Lett. 61, 1446

O'Neill, James. Prodigal genius: the life of Nikola Tesla. Book Tree, 2007.

Pais, Abraham. Subtle is the Lord: The Science and the Life of Albert Einstein: The

Science and the Life of Albert Einstein. Oxford University Press, USA, 1982.

K. Schwarzschild, "On the Gravitational Field of a Point Mass in Einstein's Theory,"

Proceedings of the Prussian Academy of Sciences, 424 (1916)

Seifer, Marc. Wizard: The life and times of Nikola Tesla. Citadel, 1998.

Stachel, John. Einstein from'B'to'Z'. Vol. 9. Springer Science & Business Media, 2001.

Storr, Anthony. "Isaac Newton." Br Med J (Clin Res Ed) 291.6511 (1985): 1779-1784.

H W Turnbull (ed.), Correspondence of Isaac Newton, Vol 2 (1676–1687), (Cambridge University Press, 1960), giving the Hooke-Newton correspondence (of November 1679 to January 1679/80) at pp.297–314, and the 1686 correspondence over Hooke's priority claim at pp.431–448.

Westfall, Richard S. The Life of Isaac Newton. Cambridge University Press, 1994.

Wilson, E. Bright. "Einstein and quantum mechanics." International Journal of
 Quantum Chemistry 16.S13 (1979): 1-4.

White, Michael. Isaac Newton: the last sorcerer. Da Capo Press, 1999.

Nobel Lectures, Physics 1901-1921, Elsevier Publishing Company, Amsterdam, 1967

www.ingramcontent.com/pod-product-compliance
Lightning Source LLC
Chambersburg PA
CBHW070230190526
45169CB00001B/144